Food Quality Management

Food Quality Management

Emmett Norton

Larsen & Keller
www.larsen-keller.com

Food Quality Management
Emmett Norton
ISBN: 978-1-64172-674-0 (Hardback)

Larsen & Keller
Published by Larsen and Keller Education,
5 Penn Plaza,
19th Floor,
New York, NY 10001, USA

Cataloging-in-Publication Data

Food quality management / Emmett Norton.
p. cm.
Includes bibliographical references and index.
ISBN 978-1-64172-674-0
1. Food--Quality--Management. 2. Food--Safety measures. 3. Food industry and trade--Quality control. I. Norton, Emmett
TP372.5 .F66 2022
363.192--dc23

For more information regarding Larsen and Keller Education and its products, please visit the publisher's website www.larsen-keller.com

Table of Contents

Preface

It is with great pleasure that I present this book. It has been carefully written after numerous discussions with my peers and other practitioners of the field. I would like to take this opportunity to thank my family and friends who have been extremely supporting at every step in my life.

The sum of all analyzable attributes of a food item such as its sensoric value, suitability value and health value are termed as food quality. Some of the factors which are considered under this are the flavor, texture and appearance of the food item. Food quality management is responsible for ensuring that the food items are safe and fit for consumption. There are various important aspects of food quality management such as quality assurance and quality control. Quality assurance seeks to prevent any defect in the food item by studying its processing. Quality control focuses on defect identification and deals with the product. This book presents the complex subject of food quality management in the most comprehensible and easy to understand language. The topics covered herein deal with the core subjects of this field. The book is appropriate for students seeking detailed information in this area as well as for experts.

The chapters below are organized to facilitate a comprehensive understanding of the subject:

Chapter - What is Food Safety?

The discipline that studies the preparation, handling and storage of food in order to prevent contamination of food and food borne illnesses is referred to as food safety. This is an introductory chapter which will briefly introduce the important aspects of food safety and quality.

Chapter - Food Quality

The quality characteristics of food, such as health value, appearance, texture and flavor, that are acceptable to consumers is defined as food quality. Some of the processes to ensure the quality of food items are food fortification, food grading and food sampling. This chapter has been carefully written to provide an easy understanding of these processes related to food quality.

Chapter - Methods of Food Safety

There are various methods and technologies that are used to ensure the safety of food. A few of its methods and technologies are pasteurization, pascalization, hurdle technology, food irradiation, etc. The topics elaborated in this chapter will help in gaining a better perspective about these methods of food safety.

Chapter - Preservation Techniques

The food can be preserved by using many preservation techniques such as salting, pickling, sugaring, canning, food drying, smoking, refrigeration and by using fruit preserves, jams and jellies. This chapter closely examines these key preservation techniques to provide an extensive understanding of the subject.

Chapter - Food Pathogens

The parasites in food that cause illness or food poisoning are referred to as food pathogens. Some of the plant pathogens are Escherichia coli O157:H7, Campylobacter, Clostridium perfringens, Listeria monocytogenes, Norovirus, Toxoplasmosis, Giardiasis, etc. This chapter discusses these food pathogens in detail.

Chapter - Effects of Food Contamination

The presence of harmful microorganisms and chemicals in food that can cause food-borne illness is defined as food contamination. It can cause many diseases such as diarrhea, gastro¬enteritis, nausea, etc. All these effects of food contamination have been carefully analyzed in this chapter.

Chapter - Food Allergies

An abnormal immune response to food is known as a food allergy. A few of such food allergies include corn allergy, egg allergy, fruit allergy, milk allergy, soy allergy, wheat allergy and peanut allergy. This chapter has been carefully written to provide an easy understanding of these kinds of food allergies.

Emmett Norton

1

What is Food Safety?

The discipline that studies the preparation, handling and storage of food in order to prevent contamination of food and food borne illnesses is referred to as food safety. This is an introductory chapter which will briefly introduce the important aspects of food safety and quality.

Food

Food is a substance consisting essentially of protein, carbohydrate, fat, and other nutrients used in the body of an organism to sustain growth and vital processes and to furnish energy. The absorption and utilization of food by the body is fundamental to nutrition and is facilitated by digestion. Plants, which convert solar energy to food by photosynthesis, are the primary food source. Animals that feed on plants often serve as sources of food for other animals.

Hunting and gathering, horticulture, pastoralism, and the development of agriculture are the primary means by which humans have adapted to their environments to feed themselves. Food has long served as a carrier of culture in human societies and has been a driving force for globalization. This was especially the case during the early phases of European trade and colonial expansion, when foods such as the hot red pepper, corn (maize), and sweet potatoes spread throughout Europe to Africa and Asia.

Food Safety and Quality

The terms food safety and food quality can sometimes be confusing. Food safety refers to all those hazards, whether chronic or acute, that may make food injurious to the health of the consumer. It is not negotiable. Quality includes all other attributes that influence a product's value to the consumer. This includes negative attributes such as spoilage, contamination with filth, discoloration, off-odours and positive attributes such as the origin, colour, flavour, texture and processing method of the food. This distinction between safety and quality has implications for public policy and influences the nature and content of the food control system most suited to meet predetermined national objectives.

Food control is defined as: A mandatory regulatory activity of enforcement by national or local authorities to provide consumer protection and ensure that all foods during production, handling, storage, processing, and distribution are safe, wholesome and fit for human consumption; conform to safety and quality requirements; and are honestly and accurately labelled as prescribed by law.

The foremost responsibility of food control is to enforce the food law(s) protecting the consumer against unsafe, impure and fraudulently presented food by prohibiting the sale of food not of the nature, substance or quality demanded by the purchaser.

Confidence in the safety and integrity of the food supply is an important requirement for consumers. Foodborne disease outbreaks involving agents such as Escherichia coli, Salmonella and chemical contaminants highlight problems with food safety and increase public anxiety that modern farming systems, food processing and marketing do not provide adequate safeguards for public health. Factors which contribute to potential hazards in foods include improper agricultural practices; poor hygiene at all stages of the food chain; lack of preventive controls in food processing and preparation operations; misuse of chemicals; contaminated raw materials, ingredients and water; inadequate or improper storage, etc.

Specific concerns about food hazards have usually focused on:

- Microbiological hazards;
- Pesticide residues;
- Misuse of food additives;
- Chemical contaminants, including biological toxins; and
- Adulteration.

The list has been further extended to cover genetically modified organisms, allergens, veterinary drugs residues and growth promoting hormones used in the production of animal products.

Consumers expect protection from hazards occurring along the entire food chain, from primary producer through consumer (often described as the farm-to-table continuum). Protection will only occur if all sectors in the chain operate in an integrated way, and food control systems address all stages of this chain.

As no mandatory activity of this nature can achieve its objectives fully without the cooperation and active participation of all stakeholders e.g. farmers, industry, and consumers, the term Food Control System is used in these Guidelines to describe the integration of a mandatory regulatory approach with preventive and educational strategies that protect the whole food chain. Thus an ideal food control system should include effective enforcement of mandatory requirements, along with training and education, community outreach programmes and promotion of voluntary compliance. The introduction of preventive approaches such as the Hazard Analysis Critical Control Point System (HACCP), have resulted in industry taking greater responsibility for and control of food safety risks. Such an integrated approach facilitates improved consumer protection, effectively stimulates agriculture and the food processing industry, and promotes domestic and international food trade.

Global Considerations

International Trade

With an expanding world economy, liberalization of food trade, growing consumer demand, developments in food science and technology, and improvements in transport and communication, international trade in fresh and processed food will continue to increase.

Access of countries to food export markets will continue to depend on their capacity to meet the regulatory requirements of importing countries. Creating and sustaining demand for their food products in world markets relies on building the trust and confidence of importers and consumers in the integrity of their food systems. With agricultural production the focal point of the economies of most developing countries, such food protection measures are essential.

Codex Alimentarius Commission

The Codex Alimentarius Commission (CAC) is an intergovernmental body that coordinates food standards at the international level. Its main objectives are to protect the health of consumers and ensure fair practices in food trade. The CAC has proved to be most successful in achieving international harmonization in food quality and safety requirements. It has formulated international standards for a wide range of food products and specific requirements covering pesticide residues, food additives, veterinary drug residues, hygiene, food contaminants, labelling etc. These Codex recommendations are used by governments to determine and refine policies and programmes under their national food control system. More recently, Codex has embarked on a series of activities based on risk assessment to address microbiological hazards in foods, an area previously unattended. Codex work has created worldwide awareness of food safety, quality and consumer protection issues, and has achieved international consensus on how to deal with them scientifically, through a risk-based approach. As a result, there has been a continuous appraisal of the principles of food safety and quality at the international level. There is increasing pressure for the adoption of these principles at the national level.

SPS and TBT Agreements

The conclusion of the Uruguay Round of Multilateral Trade Negotiations in Marrakech led to the establishment of the WTO on 1 January 1995, and to the coming into force of the Agreement on the Application of Sanitary and Phytosanitary Measures (SPS) and the Agreement on Technical Barriers to Trade (TBT). Both these Agreements are relevant in understanding the requirements for food protection measures at the national level, and the rules under which food is traded internationally.

The SPS Agreement confirms the right of WTO member countries to apply measures to protect human, animal and plant life and health. The Agreement covers all relevant laws, decrees, regulations; testing, inspection, certification and approval procedures; and packaging and labelling requirements directly related to food safety. Member States are asked to apply only those measures for protection that are based on scientific principles, only to the extent necessary, and not in a manner which may constitute a disguised restriction on international trade. The Agreement encourages use of international standards, guidelines or recommendations where they exist, and identifies those from Codex (relating to food additives, veterinary drugs and pesticide residues, contaminants, methods of analysis and sampling, and codes and guidelines of hygienic practices), to be consistent with provisions of SPS. Thus, the Codex standards serve as a benchmark for comparison of national sanitary and phytosanitary measures. While it is not compulsory for Member States to apply Codex Standards, it is in their best interests to harmonize their national food standards with those elaborated by Codex.

The TBT Agreement requires that technical regulations on traditional quality factors, fraudulent practices, packaging, labelling etc. imposed by countries will not be more restrictive on imported products than they are on products produced domestically. It also encourages use of international standards.

Elements of a National Food Control System

Objectives

The principal objectives of national food control systems are:

- Protecting public health by reducing the risk of foodborne illness.
- Protecting consumers from unsanitary, unwholesome, mislabelled or adulterated food.
- Contributing to economic development by maintaining consumer confidence in the food system and providing a sound regulatory foundation for domestic and international trade in food.

Scope

Food control systems should cover all food produced, processed and marketed within the country, including imported food. Such systems should have a statutory basis and be mandatory in nature.

Building Blocks

While the components and priorities of a food control system will vary from country to country, most systems will typically comprise the following components.

Food Laws and Regulations

The development of relevant and enforceable food laws and regulations is an essential component of a modern food control system. Many countries have inadequate food legislation and this will impact on the effectiveness of all food control activities carried out in the country.

Food law has traditionally consisted of legal definitions of unsafe food, and the prescription of enforcement tools for removing unsafe food from commerce and punishing responsible parties after the fact. It has generally not provided food control agencies with a clear mandate and authority to prevent food safety problems. The result has been food safety programmes that are reactive and enforcement-oriented rather than preventive and holistic in their approach to reducing the risk of foodborne illness. To the extent possible, modern food laws not only contain the necessary legal powers and prescriptions to ensure food safety, but also allow the competent food authority or authorities to build preventive approaches into the system.

In addition to legislation, governments need updated food standards. In recent years, many highly prescriptive standards have been replaced by horizontal standards that address the broad issues involved in achieving food safety objectives. While horizontal standards are a viable approach to delivering food safety goals, they require a food chain that is highly controlled and supplied with good data on food safety risks and risk management strategies and as such may not be feasible for many developing countries. Similarly, many standards on food quality issues have been cancelled and replaced by labelling requirements.

In preparing food regulations and standards, countries should take full advantage of Codex standards and food safety lessons learned in other countries. Taking into account the experiences in other countries while tailoring the information, concepts and requirements to the national context is the only sure way to develop a modern regulatory framework that will both satisfy national needs and meet the demands of the SPS Agreement and trading partners.

Food legislation should include the following aspects:

- It must provide a high level of health protection;
- It should include clear definitions to increase consistency and legal security;
- It should be based on high quality, transparent, and independent scientific advice following risk assessment, risk management and risk communication;
- It should include provision for the use of precaution and the adoption of provisional measures where an unacceptable level of risk to health has been identified and where full risk assessment could not be performed;
- It should include provisions for the right of consumers to have access to accurate and sufficient information;
- It should provide for tracing of food products and for their recall in case of problems;
- It should include clear provisions indicating that primary responsibility for food safety and quality rests with producers and processors;
- It should include obligation to ensure that only safe and fairly presented food is placed on the market;
- It should also recognise the country's international obligations particularly in relation to trade;
- It should ensure transparency in the development of food law and access to information.

Food Control Management

Effective food control systems require policy and operational coordination at the national level. While the detail of such functions will be determined by the national legislation, they would include the establishment of a leadership function and administrative structures with clearly defined accountability for issues such as: the development and implementation of an integrated national food control strategy; operation of a national food control programme; securing funds and allocating resources; setting standards and regulations; participation in international food control related activities; developing emergency response procedures; carrying out risk analysis; etc.

Core responsibilities include the establishment of regulatory measures, monitoring system performance, facilitating continuous improvement, and providing overall policy guidance.

Inspection Services

The administration and implementation of food laws require a qualified, trained, efficient and honest food inspection service. The food inspector is the key functionary who has day-to-day contact with the food industry, trade and often the public. The reputation and integrity of the food control system depends, to a very large extent, on their integrity and skill. The responsibilities of the inspection services include:

- Inspecting premises and processes for compliance with hygienic and other requirements of standards and regulations;
- Evaluating HACCP plans and their implementation;
- Sampling food during harvest, processing, storage, transport, or sale to establish compliance, to contribute data for risk assessments and to identify offenders;
- Recognizing different forms of food decomposition by organoleptic assessment ; identifying food which is unfit for human consumption; or food which is otherwise deceptively sold to the consumer; and taking the necessary remedial action;
- Recognizing, collecting and transmitting evidence when breaches of law occur, and appearing in court to assist prosecution;
- Encouraging voluntary compliance in particular by means of quality assurance procedures;
- Carrying out inspection, sampling and certification of food for import/export inspection purposes when so required;
- In establishments working under safety assurance programmes such as HACCP, conduct risk based audits.

Proper training of food inspectors is a prerequisite for an efficient food control system. As current food systems are quite complex, the food inspector must be trained in food science and technology to understand the industrial processes, identify potential safety and quality problems, and have the skill and experience to inspect the premises, collect food samples and carry out an overall evaluation. The inspector must have a good understanding of the relevant food laws and regulations, their powers under those laws, and the obligations such laws impose on the food sector. They should also be conversant with procedures for collecting evidence, writing inspection reports, collecting samples and sending them to a laboratory for analysis. With gradual introduction of HACCP systems in the food industry, the inspector should be trained to handle HACCP audit responsibilities. Clearly, there is a continuing need for training and upgrading the skills of existing inspectional staff and having a policy for human resource development, especially the development of inspectional specialists in specific technical areas.

As human resources in some food control agencies in developing countries may be limited, environmental health inspectors are often also asked to work as food inspectors. This is not the ideal situation as they may lack the skills and knowledge to effectively evaluate and inspect food operations. If environmental health inspectors must be used, then they should be carefully supervised and provided with on-the-job training.

Laboratory Services: Food Monitoring and Epidemiological Data

Laboratories are an essential component of a food control system. The establishment of laboratories requires considerable capital investment and they are expensive to maintain and operate. Therefore careful planning is necessary to achieve optimum results. The number and location of the laboratories should be determined in relation to the objectives of the system and the volume of work. If more than one laboratory is required, consideration should be given to apportioning the analytical work to achieve the most effective coverage of the food analyses to be performed and also to having a central reference laboratory equipped for sophisticated and reference analyses.

All food analysis laboratories may not be under the control of one agency or ministry, and a number could be under the jurisdiction of the states, provinces and local authorities. The Food Control Management should, however, lay down the norms for food control laboratories and monitor their performance.

The laboratories should have adequate facilities for physical, microbiological and chemical analyses. In addition to simple routine analysis, the laboratories can be equipped with more sophisticated instruments, apparatus and library facilities as required. It is not only the type of equipment that determines the accuracy and reliability of analytical results but also the qualification and skill of the analyst and the reliability of the method used. The analytical results of a food control laboratory are often used as evidence in a court of law to determine compliance with regulations or standards of the country. It is therefore necessary that utmost care be taken to ensure the efficient and effective performance of the laboratory. The introduction of analytical quality assurance programmes and accreditation of the laboratory by an appropriate accreditation agency within the country or from outside, enables the laboratory to improve its performance and to ensure reliability, accuracy and repeatability of its results. Prescription of official methods of sampling and analysis also support this effort.

An important element of a national food control system is its integration in a national food safety system so that links between food contamination and foodborne diseases can be established and analyzed. Access to reliable and current intelligence on the incidence of foodborne illness is critical. The laboratory facilities for this type of activity are generally situated outside the food control agencies. It is essential, however, that effective linkages are established between food control agencies and the public health system including epidemiologists and microbiologists. In this way information on foodborne diseases may be linked with food monitoring data, and lead to appropriate risk-based food control policies. This information includes annual incidence trends, identification of susceptible population groups, identification of hazardous foods, identification and tracing of causes of foodborne diseases, and the development of early warning systems for outbreaks and food contamination.

Information, Education, Communication and Training

An increasingly important role for food control systems is the delivery of information, education and advice to stakeholders across the farm-to-table continuum. These activities include the provision of balanced factual information to consumers; the provision of information packages and educational programmes for key officials and workers in the food industry; development of train-the-trainer programmes; and provision of reference literature to extension workers in the agriculture and health sectors.

Food control agencies should address the specific training needs of their food inspectors and laboratory analysts as a high priority. These activities provide an important means of building food control expertise and skills in all interested parties, and thereby serve an essential preventive function.

Strengthening National Food Control Systems

Principles of Food Control: Issues for Consideration

When seeking to establish, update, strengthen or otherwise revise food control systems, national authorities must take into consideration a number of principles and values that underpin food control activities, including the following:

- Maximizing risk reduction by applying the principle of prevention as fully as possible throughout the food chain;
- Addressing the farm-to-table continuum;
- Establishing emergency procedures for dealing with particular hazards (e.g. recall of products);
- Developing science-based food control strategies;
- Establishing priorities based on risk analysis and efficacy in risk management;
- Establishing holistic, integrated initiatives which target risks and impact on economic wellbeing;
- Recognizing that food control is a widely shared responsibility that requires positive interaction between all stakeholders.

Certain key principles and related issues are discussed below:

Integrated Farm-to-table Concept

The objective of reduced risk can be achieved most effectively by the principle of prevention throughout the production, processing and marketing chain. To achieve maximum consumer protection it is essential that safety and quality be built into food products from production through to consumption. This calls for a comprehensive and integrated farm-to-table approach in which the producer, processor, transporter, vendor, and consumer all play a vital role in ensuring food safety and quality.

It is impossible to provide adequate protection to the consumer by merely sampling and analysing the final product. The introduction of preventive measures at all stages of the food production and distribution chain, rather than only inspection and rejection at the final stage, makes better economic sense, because unsuitable products can be identified earlier along the chain. The more economic and effective strategy is to entrust food producers and operators with primary responsibility for food safety and quality. Government regulators are then responsible for auditing performance of the food system through monitoring and surveillance activities and for enforcing legal and regulatory requirements.

Food hazards and quality loss may occur at a variety of points in the food chain, and it is difficult and expensive to test for their presence. A well structured, preventive approach that controls processes is the preferred method for improving food safety and quality. Many but not all potential food hazards can be controlled along the food chain through the application of good practices i.e. good agricultural practices (GAP), good manufacturing practices (GMP), and good hygienic practices (GHP).

An important preventative approach that may be applied at all stages in the production, processing and handling of food products involves the Hazard Analysis Critical Control Point system (HACCP). The principles of HACCP have been formalised by the Codex Committee on Food Hygiene1 , and provide a systematic structure to the identification and control of foodborne hazards. Governments should recognize the application of a HACCP approach by the food industry as a fundamental tool for improving the safety of food.

Risk Analysis

The Codex Alimentarius Commission defines risk analysis as a process composed of three components:

- Risk assessment - A scientifically based process consisting of the following steps: (i) hazard identification; (ii) hazard characterization; (iii) exposure assessment; and (iv) risk characterization.
- Risk management - The process, distinct from risk assessment, of weighing policy alternatives, in consultation with all interested parties, considering risk assessment and other factors relevant for the health protection of consumers and for the promotion of fair trade practices, and, if needed selecting appropriate prevention and control options.
- Risk communication - The interactive exchange of information and opinions throughout the risk analysis process concerning hazards and risks, risk related factors and risk perceptions, among risk assessors, risk managers, consumers, industry, the academic community and other interested parties, including the explanation of risk assessment findings and the basis of risk management decisions.

Risk analysis is well established for chemical hazards, and FAO and WHO are now extending the experience and expertise developed from risk analysis of chemical hazards to that of microbiological hazards.

Risk analysis must be the foundation on which food control policy and consumer protection measures are based. While not all countries may have sufficient scientific resources, capabilities, or data to carry out risk assessments, it may not even be necessary in all cases to generate local data for this purpose. Instead countries should make full use of the international data and expertise as well as data from other countries that are consistent with internationally accepted approaches. Risk assessments carried out at the international level by JECFA, JMPR, and other expert bodies are particularly useful. Developing countries should take a pragmatic approach and develop a cadre of scientists to interpret such data and assessments, and to use this information for the development of national food control programmes.

Codex standards take into account risk assessments carried out at the international level and are accepted as scientifically valid under the SPS Agreement. Hence their adoption and implementation within national food control systems is encouraged.

Risk management should take into account the economic consequences and feasibility of risk management options, and recognize the need for flexibility consistent with consumer protection requirements.

Transparency

A food control system must be developed and implemented in a transparent manner. The confidence of consumers in the safety and quality of the food supply depends on their perception of the integrity and effectiveness of food control operations and activities. Accordingly, it is important that all decision-making processes are transparent, allow all stakeholders in the food chain to make effective contributions, and explain the basis for all decisions. This will encourage cooperation from all concerned parties and improve the efficiency and rate of compliance.

Food control authorities should also examine the manner in which they communicate food safety information to the public. This may take the form of scientific opinion on food safety matters, overviews of inspection activity, and findings on foods implicated in foodborne illnesses, food poisoning episodes, or gross adulteration. All this could be considered as a part of risk communication to enable consumers to better understand the risks and their responsibilities for minimizing the impact of foodborne hazards.

Regulatory Impact Assessment

When planning and implementing food control measures, consideration must be given to the costs of compliance (resources, personnel, and financial implications) to the food industry, as these costs are ultimately passed onto consumers. The important questions are: Do the benefits of regulation justify the costs? What is the most efficient management option? Export inspection systems designed to assure the safety and quality of exported foods, will protect international markets, generate business and secure returns. Animal and plant health measures improve agricultural productivity. In contrast, food safety is an essential public health goal and may impose costs on producers, yet investments in food safety may not be immediately rewarded in the market place.

Regulatory impact assessments (RIA) are of increasing importance in determining priorities and assist food control agencies in adjusting or revising their strategies to achieve the most beneficial effect. They are, however, difficult to carry out. Two approaches have been suggested for determining cost/benefit of regulatory measures in food safety:

- Theoretical models can be developed to estimate willingness to pay (WTP) for reduced risk of morbidity and mortality; and
- Cost of illness (COI) covering lifetime medical costs and lost productivity.

Both approaches require considerable data for interpretation. COI estimates are perhaps easier for policy makers to understand and have been widely used to justify measures for food control, even though they do not measure the full value of risk reduction. Not surprisingly, it is easier to perform a RIA for an export inspection intervention, than for regulatory policy which achieves a public health outcome.

Developing a National Food Control Strategy

The attainment of food control system objectives requires knowledge of the current situation and the development of a national food control strategy. Programmes to achieve these objectives tend to be country specific. Like socioeconomic considerations, they are also influenced by current or emerging food safety and quality issues. Such programmes also need to consider international perceptions of food risks, international standards, and any international commitments in the food protection area. Therefore, when establishing a food control system it is necessary to systematically examine all factors that may impinge upon the objectives and performance of the system, and develop a national strategy.

Collection of Information

This is achieved through the collection and collation of relevant data in the form of a Country Profile. This data underpins strategy development, with stakeholders reaching consensus on objectives, priorities, policies, roles of different ministries/agencies, industry responsibilities, and timeframe for implementation. In particular, major problems associated with the control and prevention of foodborne diseases are identified so that effective strategies for the resolution of these problems can be implemented.

The profile should permit a review of health and socioeconomic issues impacting on foodborne hazards, consumers concerns, and the growth of industry and trade, as well as identification of the functions of all sectors which are directly and indirectly involved in ensuring food safety and quality and consumer protection. The collection of epidemiological data on foodborne illness is an indispensable component of a country profile and should be done whenever possible.

Development of Strategy

The preparation of a national food control strategy enables the country to develop an integrated, coherent, effective and dynamic food control system, and to determine priorities which ensure consumer protection and promote the country's economic development. Such a strategy should provide better coherence in situations where there are several food control agencies involved with no existing national policy or overall coordinating mechanism. In such cases, it prevents confusion, duplication of effort, inefficiencies in performance, and wastage of resources.

Devising strategies for food control with clearly defined objectives is not simple, and the identification of priorities for public investment in food control can be a challenging task. The strategy should be based on multi-sectoral inputs and focus on the need for food security, and consumer protection from unsafe adulterated or misbranded food. At the same time it should take into consideration the economic interests of the country in regard to export/import trade, the development of the food industry, and the interests of farmers and food producers. Strategies should use a risk based approach to determine priorities for action. Areas for voluntary compliance and mandatory action should be clearly identified,and timeframes determined. The need for human resource development and strengthening of infrastructure such as laboratories should be also considered.

Certain types of food control interventions require large fixed capital investments in equipment and human resources. While it is easier to justify these costs for larger enterprises, imposing such costs on smaller firms who may coexist with larger enterprises may not be appropriate. Therefore the gradual phasing in of such interventions is desirable. For example, countries may allow small enterprises longer periods of time to introduce HACCP.

The strategy will be influenced by the country's stage of development, the size of its economy, and the level of sophistication of its food industry. The final strategy should include:

- A national strategy for food control with defined objectives, a plan of action for its implementation, and milestones;
- Development of appropriate food legislation, or revision of the existing legislation to achieve the objectives defined by the national strategy;
- Development or revision of food regulations, standards and codes of practice as well as harmonizing these with international requirements;
- A programme for strengthening food surveillance and control systems;
- Promotion of systems for improving food safety and quality along the food chain i.e. introduction of HACCP-based food control programmes;
- Development and organization of training programmes for food handlers and processors, food inspectors, and analysts;
- Enhanced inputs into research, foodborne disease surveillance, and data collection, as well as creating increased scientific capacity within the system; and
- Promotion of consumer education and other community outreach initiatives.

Strengthening Organizational Structures for National Food Control Systems

Given the wide scope of food control systems, there are at least three types of organizational arrangements that may be appropriate at the national level. These are:

- A system based on multiple agencies responsible for food control - Multiple Agency System;
- A system based on a single, unified agency for food control - Single Agency System;
- A system based on a national integrated approach - Integrated System.

Multiple Agency System

While food safety is the foremost objective, food control systems also have an important economic objective of creating and maintaining sustainable food production and processing systems. In this context, food control systems play a significant role in the following:

- Ensuring fair practices in trade;
- Developing the food sector on a professional and scientific basis;
- Preventing avoidable losses and conserving natural resources; and
- Promoting the country's export trade.

The systems that deal specifically with these objectives can be sectoral i.e. based upon the need for development of the particular sector such as fisheries, meat and meat products, fruit and vegetables, milk and milk products. These systems can be mandatory or voluntary, and put into effect either through a general food law or a sectoral regulation. Examples include:

- An export inspection law that identifies foods to be covered for mandatory export inspection prior to export; or offers facilities for voluntary inspection and certification for exporters.
- Specific commodity inspection regulations, such as for fish and fish products, meat and meat products, or fruit and vegetable products which are implemented by different agencies or ministries given this mandate under relevant law(s).
- Regulated systems for grading and marking of fresh agricultural produce which go directly for sale to the consumer or as raw material for industry. They are mostly confined to quality characteristics so that the producer gets a fair return for his produce and the buyer is not cheated.

Where sectoral initiatives have resulted in the establishment of separate food control activities, the outcome has been the creation of multiple agencies with responsibilities for food control. Typically, under such arrangements the food control responsibilities are shared between Government Ministries such as Health, Agriculture, Commerce, Environment, Trade and Industry, and Tourism, and the roles and responsibilities of each of these agencies are specified but quite different. This sometimes leads to problems such as duplication of regulatory activity, increased bureaucracy, fragmentation, and a lack of coordination between the different bodies involved in food policy, monitoring, and control of food safety. For example, the regulation and surveillance of meat and meat products may be separate from food control undertaken by a Ministry of Health. Meat inspection is often done by Ministry of Agriculture or primary industry personnel who undertake all veterinary activities, and the data generated may not be linked to public health and food safety monitoring programmes.

Food control systems may also be fragmented between national, state and local bodies, and the thoroughness of implementation depends upon the capacity and the efficiency of the agency responsible at each level. Thus consumers may not receive the same level of protection throughout the country and it may become difficult to properly evaluate the effectiveness of interventions by national, state or local authorities.

While multiple food control agencies may be the norm, they suffer from serious drawbacks including:

- Lack of overall coordination at national level;
- Frequent confusion over jurisdiction and resultant inefficiencies in performance;
- Differences in levels of expertise and resources and hence uneven implementation;
- Conflict between public health objectives and the facilitation of trade and industry development;
- Limited capacity for appropriate scientific inputs in decision-making processes;
- Lack of coherence leading to over-regulation or time gaps in adequate regulatory activity; and
- Reductions in the confidence of domestic consumers and foreign buyers in the credibility of the system.

During the preparation of a national food control strategy, it is important to consider the type and size of the organization(s) that are necessary to implement the strategy. It is often not possible to have a single unified structure or an integrated food control system, due to various historical and political reasons. In such cases, it is necessary for the national food control strategy to clearly identify the role of each agency to avoid duplication of effort and to bring about a measure of coherence among them. It should also identify areas or segments of the food chain which require special attention and need additional resources for strengthening.

Single Agency System

The consolidation of all responsibility for protecting public health and food safety into a single food control agency with clearly defined terms of reference has considerable merit. It acknowledges the high priority that Government places in food safety initiatives and a commitment to reducing the risk of foodborne disease. The benefits that result from a single agency approach to food control include:

- Uniform application of protection measures;
- Ability to act quickly to protect consumers;
- Improved cost efficiency and more effective use of resources and expertise;
- Harmonization of food standards;
- Capacity to quickly respond to emerging challenges and the demands of the domestic and international marketplace; and
- The provision of more streamlined and efficient services, benefiting industry and promoting trade.

While a national strategy helps to influence both the legislation and the organizational structure for enforcement, it is not possible to recommend a single organizational structure that will universally meet the requirements and resources of every country's socioeconomic and political environment. The decision has to be country specific and all stakeholders should have the opportunity to provide inputs into the development process. Unfortunately, there are often few opportunities for countries to build a new food control system based on a single agency.

Integrated System

Integrated food control systems warrant consideration where there is desire and determination to achieve effective collaboration and coordination between agencies across the farm-to-table continuum. Typically, the organization of an integrated food control system would have several levels of operation:

Level 1: Formulation of policy, risk assessment and management, and development of standards and regulations.

Level 2: Coordination of food control activity, monitoring, and auditing.

Level 3: Inspection, and enforcement.

Level 4: Education and training.

In reviewing and revising their food control systems, governments may wish to consider a model which calls for the establishment of an autonomous national food agency which is responsible for activities at Levels 1 and 2, with existing multi-sectoral agencies retaining responsibility for Level 3 and 4 activities. The advantages of such a system include:

- Provides coherence in the national food control system;
- Politically more acceptable as it does not disturb the day to day inspection and enforcement role of other agencies;
- Promotes uniform application of control measures across the whole food chain throughout the country;
- Separates risk assessment and risk management functions, resulting in objective consumer protection measures with resultant confidence among domestic consumers and credibility with foreign buyers;
- Better equipped to deal with international dimensions of food control such as participation in work of Codex, follow-up on SPS/TBT Agreements, etc;
- Encourages transparency in decision-making processes, and accountability in implementation;
- Is more cost-effective in the long term.

Responding to these benefits, several countries have established or are in the process of creating such a policy making and coordinating mechanism at the national level.

By placing management of the food supply chain under a competent, autonomous agency, it is possible to fundamentally change the way food control is managed. The role of such an agency is to establish national food control goals, and put into effect the strategic and operational activities necessary to achieve those goals. Other functions of such a body at the national level may include:

- Revising and updating the national food control strategy as needed;
- Advising relevant ministerial officials on policy matters, including determination of priorities and use of resources;
- Drafting regulations, standards and codes of practice and promoting their implementation;
- Coordinating the activity of the various inspection agencies, and monitoring performance;
- Developing consumer education and community outreach initiatives and promoting their implementation;
- Supporting research and development;
- Establishing quality assurance schemes for industry and supporting their implementation.

An integrated National Food Control Agency should address the entire food chain from farm-to-table, and should have the mandate to move resources to high priority areas and to address important sources of risk. The establishment of such an agency should not involve day-to-day food inspection responsibilities. These should continue to lie with existing agencies at national, state/provincial, and local levels. The agency should also consider the role of private analytical, inspection, and certification services particularly for export trade.

Funding National Food Control Systems

The funds and resources required for reorganizing and strengthening food control systems would normally be made available from the national government. In countries where food control responsibilities are spread across many government agencies it may be necessary to negotiate a revised funding structure and establish transition arrangements to ensure continuity of funds and resources. For this to occur, it is essential there is full commitment by the government for establishing appropriate structures and developing policies to deliver the optimum level of consumer protection.

Securing sufficient resources may be a problem, as the trend towards reduced public sector spending is influencing governments to review their priorities and funding arrangements. Cost recovery is practised in many countries. It is important that this is managed carefully as any costs passed directly onto the food industry will ultimately be passed onto consumers as an indirect tax on food. This falls disproportionately on the poorer sectors of society. Cost recovery options include fees for licensing, inspection activity, and food analysis. In some countries the trend towards smaller governments has resulted in the contracting of food control services from the private sector. This involves private providers contracted to undertake specific food control activities such as food inspection and surveillance.

Specific Issues of Developing Countries

Food Systems

Food production, processing, and marketing systems are complex. In many developing countries they are also highly fragmented and dependent upon a large number of small producers. While this may have socioeconomic benefits, as large quantities of food pass through a multitude of food handlers and middlemen, the risk of exposing food to unhygienic environments, contamination and adulteration increases. Problems occur as a result of poor postharvest handling, processing and storage of food and also due to inadequate facilities and infrastructure such as the absence or shortage of safe water supply, electricity, storage facilities including cold stores, and transport facilities and networks, etc. Furthermore, a majority of food producers and handlers lack appropriate knowledge and expertise in the application of modern agricultural practices, food hygiene, and good food handling practices.

This does not mean that all food from such sources is unsafe. Many traditional food production and handling practices have in-built food safety margins based on years of experience. Problems arise because of the inability to cope with the introduction of emerging intensive agricultural practices, increasing urbanization, stress on natural resources, and new food safety risks.

Food Processing Industry

The food processing industry in developing countries ranges from sophisticated state-of-theart facilities to small artisanal operations producing traditional foods for the local community. The size of these processing units is quite variable – from a few large plants to a majority of small and cottage scale units with very limited resources for effective technological inputs. At the least developed end of this continuum, these premises are ill equipped to deal with the maintenance of food safety and quality in a scientific and sustained manner. Governments often support these small units as they provide employment and generate income for their operators. The challenge for developing countries is to provide incentives for the effective expansion of these small units so they may absorb better technology.

Food processors in developing countries also face problems with the reliability and timely delivery of raw material, as well as variations in overall quality. Small holders usually produce raw materials, and a lack of infrastructure in the producing areas results in variability in the quality of these materials. This calls for greater vigilance by the food processing units and for food control activity to be implemented at all stages along the food supply chain.

Street Foods

Studies in developing countries have shown that up to 20-25% of household food expenditure is incurred outside the home, and some segments of the population depend entirely on street foods. This has been one of the consequences of rapid urbanization, with millions of people having no access to a kitchen or other cooking facilities. There are millions of single workers without families and a large floating population who move in and out of the city for work, and these people largely depend upon street foods for their daily sustenance.

In many developing countries, street food vendors are an important component of the food supply chain. Being reasonably priced and conveniently available, street food satisfies a vital need of the urban population. These ready-to-eat foods and beverages are prepared and sold by vendors or hawkers mainly in streets or other convenient public places such as around places of work, schools, hospitals, railway stations, and bus terminals.

Food safety is a major concern with street foods. These foods are generally prepared and sold under unhygienic conditions, with limited access to safe water, sanitary services, or garbage disposal facilities. Hence street foods pose a high risk of food poisoning due to microbial contamination, as well as improper use of food additives, adulteration and environmental contamination.

Food Control Infrastructure and Resources

Food control infrastructure in many developing countries tends to be inadequate, due to limited resources and often poor management. Food control laboratories are frequently poorly equipped and lack suitably trained analytical staff. This is accentuated where multiple agencies are involved in food control. A lack of overall strategic direction means that limited resources are not properly utilized. Food control systems may also suffer from poorly or inadequately developed compliance policies.

Modern food control systems call for science-based and transparent decision-making processes, and require access to qualified and trained personnel in disciplines such as food science and technology, chemistry, biochemistry, microbiology, veterinary science, medicine, epidemiology, agricultural sciences, quality assurance, auditing and food law. Food control authorities need to better appreciate the role of science in the risk-based approach, and to take advantage of scientific resources in the international community.

Technical Assistance: Role of International Agencies

The need for technical assistance in strengthening food control systems in developing countries is well recognized. FAO and WHO are the two main specialized agencies of the United Nations involved in food quality and safety technical cooperation programmes with developing countries.

FAO assistance in food control and food standards is a major activity and is delivered at global, regional, and country levels. Published manuals of food quality control cover a range of different aspects of food control systems and are used internationally. Meetings, seminars and workshops are conducted in all regions of Africa, Asia and the Pacific, Latin America and the Caribbean, Eastern Europe, the Near East and North Africa. Technical assistance is provided in many areas such as the following:

- Establishing or strengthening national food control systems and infrastructure;
- Assistance in preparation of food law and regulations;
- Workshops on developing national strategies for food control;
- Assistance in establishing or improving food analysis capabilities;
- Assessing the implications of SPS and TBT Agreements;
- Providing training in food inspection, analysis and food handling;
- Providing training of trainers in HACCP;
- Providing training in management of food control systems; and
- Assistance in strengthening National Codex Committees.

WHO has in recent years substantially increased the priority of its food safety activities at international and regional level. The Organization also provides technical assistance at international, regional, and country levels. Under its decentralized structure, WHO is divided into six regions, with Regional Offices responsible for providing assistance to Member States in developing and strengthening their National Food Safety Programmes. Regional Offices currently undertake a range of capacity building initiatives designed to safeguard consumer health. The nature and extent of these activities is influenced by available resources, but includes the following:

- Developing regional and national food safety policy and strategies;
- Preparation of food legislation, food regulations and standards, and codes of hygienic practice;
- Implementation of food inspection programmes;
- Promoting methods and technologies designed to prevent foodborne diseases, including the application of the HACCP system;
- Developing or enhancing food analysis capability;
- Development and delivery of hygiene training and education programmes;
- Establishing healthy markets and enhancing the safety of street food;
- Promoting the establishment of foodborne disease surveillance activity.

Both the SPS Agreement and TBT Agreement specifically refer to the need to provide technical assistance to developing countries. Such assistance may be in areas of processing technologies, research and infrastructure, establishment of national regulatory bodies, etc. In particular, developed countries which import food from developing nations are required, upon request, to provide technical assistance to the developing exporting countries to enable these countries to meet their SPS or TBT obligations in international food trade. This new opportunity to access technical assistance under the WTO Agreements has not yet been fully utilized by developing countries.

Technical assistance in the food control area may also be obtained through the World Bank, other development banks, and from bilateral donor agencies. Access to such funds is dependent upon the priority that developing countries attach to strengthening their food control systems as reflected in their national development plans.

ISO 22000

ISO 22000 is a Food Safety Management System that can be applied to any organization in the food chain, farm to fork. Becoming certified to ISO 22000 allows a company to show their customers that they have a food safety management system in place. This provides customer confidence in the product. This is becoming more and more important as customers demand safe food and food processors require that ingredients obtained from their suppliers to be safe.

The International Organization for Standardization (ISO) developed the Food Safety Management System Certification: ISO 22000. ISO and its member countries used the Quality Management System approach, and tailored it to apply to Food Safety, incorporating the widely used and proven HACCP principles and Good Manufacturing Principles (addressed by Prerequisite Programs in ISO 22000).

The standard has requirements for Food Safety Management Systems processes and procedures, and requires that the organization implement prerequisite programs and HACCP.

Unlike some of the other Food Safety Management Systems Certification programs (for example FSSC 22000 and SQF) the ISO 22000 does not have specific requirements for prerequisite programs (PRPs), but requires that the organization identifies and implements the appropriate programs. This makes it more flexible, and food organizations of any type can implement and be certified to ISO 22000.

Food processors and manufacturers can use the ISO Technical specification ISO/TS 22002-1 to develop their PRP programs. It outlines the requirements for PRP programs that are applicable to these organizations. The requirements outlined are widely accepted and are equivalent to the requirements in the PAS 220, the publicly available specification used along with ISO 22000 for the FSSC 22000 Certification scheme.

ISO 22000 is not a Global Food Safety Initiative (GFSI) benchmarked standard. This means that if your customer base or market is looking for a GFSI Recognized standard you should look at FSSC 22000, which is the most similar to ISO 22000 or one of the other GFSI recognized certification schemes.

What does ISO 22000 require?

ISO 22000 requires that you build a Food Safety Management System. This means that you will have a documented system in place and fully implemented throughout your facility that includes:

- Effective Prerequisite Programs in place to ensure a clean sanitary environment.
- A Hazard Analysis and Critical Control Plan developed to identify, prevent and eliminate food safety hazards.
- Established documented food safety management system processes to manage food safety throughout your organization - from management and business planning aspects to day to day communication and operations affecting food safety.

The ISO 22000 standard contains the specific requirements to be addressed by the Food Safety Management System. The standard requires food safety management system processes including:

- Having an overall Food Safety Policy for your organization, developed by top management.
- Setting objectives that will drive your companies efforts to comply with this policy.
- Planning and designing a management system and documenting the system.
- Maintaining records of the performance of the system.

- Establishing a group of qualified individuals to make up a Food Safety Team.
- Defining communication procedures to ensure effective communication with important contacts outside the company (regulatory, customers, suppliers and others) and for effective internal communication.
- Having an emergency plan.
- Holding management review meetings to evaluate the performance of the FSMS.
- Providing adequate resources for the effective operation of the FSMS including appropriately trained and qualified personnel, sufficient infrastructure and appropriate work environment to ensure food safety.
- Implementing Prerequisite Programs.
- Following HACCP principles.
- Establishing a traceability system for identification of product.
- Establishing a corrective action system and control of nonconforming product.
- Maintaining a documented procedure for handling withdrawal of product.
- Controlling monitoring and measuring devices.
- Establishing and maintaining and internal audit program.
- Continually updating and improving the FSMS.

2
Food Quality

The quality characteristics of food, such as health value, appearance, texture and flavor, that are acceptable to consumers is defined as food quality. Some of the processes to ensure the quality of food items are food fortification, food grading and food sampling. This chapter has been carefully written to provide an easy understanding of these processes related to food quality.

Food quality represents the sum of all properties and assessable attributes of a food item. Usually this is done by the three accepted categories of quality: sensoric value, suitability value and health value. All three deal with assessments, that is, judgements with a subjective component.

In addition to the value-related interpretation of quality there is the value-neutral term in the sense of condition, that is the sum of properties of a product. From this can be concluded that quality is not easily definable scientifically and that it comprises many different aspects. Obligatory and uniform definitions are also made difficult, since those aspects are subject to constant change.

Contradictions in the discussion about food quality arise mainly because of self-serving interests of producers, processors and traders of food as well as consumers, since concerning the assessment of simple quality features of products these interest groups often hold quite different views. The existing contradictions can be overcome, if all justified interests are considered, that is, with a holistic view of all the separate aspects.

A holistic assessment of quality of food comprises, in addition to the three recognised partial qualities, additional categories of quality which are currently gaining in significance. On the one hand there is a psychological or notional value of food, based on usually difficult-to-explain conceptions, opinions (prejudices) and expectations of consumers concerning a product. Foods are imputed to have certain properties which determine, whether these will be selected and eaten. Without a clear delimitation to this area foods have a cultural or social value. The prestige value of food is determined by food habits of certain population groups as well as by supply and price. Foods that are taboo or that are used as reward get their social value in this manner. The political value of foods comprises aspects like the import of foods and feeds, especially from developing countries as well as production and handling of food surplus and employment of food aid. A further category of quality is the ecological value of foods which assesses the consequences on the environment due to food production and food processing, as well as their manifold interactions and feedbacks.

These additional criteria or properties of food are often more difficult to define and to include, since they cannot be identified and measured on the product itself. From this, however, it should not be concluded that these criteria in the spectrum of food quality are not important. Even though single interest groups in the food sector will still get their way in regard to expectations and demands concerning quality, social demands and necessities are gaining increasing importance. Future and social requirements relating to the quality of food are expected to avoid misjudgements by using a holistic assessment.

Factors affecting Quality of Food

Biology and Genetics

Obviously, the quality of any muscle food depends first and foremost on the genetics and biology of the animal. The beef from a young animal is more tender than that from an old animal. Due primarily to biological reasons, muscle from some parts of beef cattle is tastier and more tender than those from another part. Chickens are more tender than turkey. White meat is biologically different from dark meat. Of course, the preference of a consumer varies with regard to the two different kinds of meat. Saltwater fish is different from freshwater fish. Some fish have more bones than others. Western consumers prefer fish with fewer bones while most often the opposite is true for Asians.

Nutrition

Recently, the nutrition of food has reached an alltime high as far as its impact on our health is concerned. There is no doubt the majority of Americans consider a quality food as one with high nutritional value. Some salient points follow:

1. Meat and poultry are nutritious because of their high source of protein, vitamins, and minerals.
2. The high content of fat and cholesterol in land muscle foods is undesirable.
3. Fish and shellfish are an important part of a healthy diet. Fish and shellfish contain high-quality protein and other essential nutrients, are low in saturated fat, and contain omega-3 fatty acids.

A well-balanced diet that includes a variety of fish and shellfish can contribute to heart health and children's proper growth and development.

Flavors and Aroma

One major reason, among many, that we like to eat is because food tastes good, which equates to flavor and aroma. Extensive research over the past 25 to 30 years has identified more than 1,000 flavor compounds in meats. However, a single compound or group of compounds responsible for "meaty flavor" has not and perhaps never will be identified due to the overall complexity of meat flavor. Meat flavor is dependent on the pool of flavor precursors in the meat tissue and the chemical reactions that occur during processing. Processing and subsequent storage contribute to the development of the characteristic flavors of meats.

Because the precise flavor precursors vary between and within species, beef, pork, lamb, and poultry each have distinctive flavor characteristics. The quality of meat and poultry is to a large extent defined by its flavor and aroma. In general, fresh saltwater fish are almost odorless because they contain a small quantity of volatiles while freshwater fish give off pyrrolidine and other earthy-odor compounds.

The compounds responsible for the development of flavor during seafood cooking can be classified in two groups. One, which represents the pleasant cucumbed green, almondnutty, and potato aroma notes, consists of highly volatile, low molecular weight compounds belonging to various chemical classes such as aldehydes, ketones, alcohols, esters, nitrogen,phenols, and sulfur-containing compounds. The second is due to water soluble, low molecular weight free amino acids (taurine, glutamic acid, glycine), nucleotides (purine derivatives), organic acids (lactic acid), and inorganic salts (Na, K, Cl).

Biogenic amines are nitrogen-containing compounds, which are present at very low levels in fresh fish. However, during storage and deterioration, biogenic amines can be produced by amino acid decarboxylation from bacterial enzymes. Among biogenic amines formed, putrescine and cadaverine have a putrid flavor while histamine and phenylethylamine have a pungent and fishy flavor, respectively. Biogenic amines are thermally stable and, therefore, have been used as indices to determine fish freshness. Volatile amines such as trimethylamine (TMA) or dimethylamine (DMA) are formed from trimethylamine oxide (TMAO), and these compounds also serve as a quality index for marine fish.

Color

The first impression that a consumer receives concerning a food product is established visually, and among the properties observed are color, form, and surface characteristics. Color is the main aspect that defines a food's quality, and a product may be rejected simply because of its color, even before other properties, such as aroma, texture, and taste, can be evaluated. This is why the appearance (optical properties, physical form, and presentation) of meat and poultry products at the point of sale is of such importance for the industry. Regarding the specific characteristics that contribute to the physical appearance of meat and poultry, color is the quality that most influences consumer choice.

Food technologists have a special interest in the color of food for several reasons. First, because of the need to maintain a uniform color throughout processing; second, to prevent any external or internal agent from acting on the product during processing, storage, and display; third, to improve or optimize a product's color and appearance; and, last, to attempt to bring the product's color into line with what the consumer expects. Put simply, the color of meat is determined by the pigments present.

These can be classified into the following types:

- Biological (carotenes and haemopigments), which are accumulated or synthesized in the organism ante-mortem.
- Pigments produced as a result of damage during manipulation or inadequate processing conditions.
- Pigments produced postmortem (through enzymatic or nonenzymatic reactions).

Those resulting from the addition of natural or artificial colorants As a quality parameter, color has been widely studied in fresh-meat and cooked products. Dry-cured meat products have received less attention because in this type of product, color formation takes place during the different processing stages. Recently, new haempigment has been identified in this type of product. From a practical point of view, color plays a fundamental role in the animal production sector, especially in meat production (primarily beef and poultry) since in many countries of the European Union, paleness receives a wholesale premium.

Microbiology and Safety

All foods contain microorganisms, some beneficial to and some with potential harm for mankind. With muscle foods, the beneficial ones are responsible for fermented meat and fish. Those potential pathogens are of concern. In the last 25 years, government records show that pathogenic organisms in meat, poultry, and seafood have been responsible for many deaths and injuries. Also, marine toxins pose big threats to our well-being considering that most of us enjoy eating fish and shellfish. It is not surprising that a quality muscle food must also be a safe one. In view of potential hazards from the consumption of muscle foods, state and federal agencies have developed and implemented stringent safety requirements in the processing of meat, poultry, and seafood.

Food Processing

The quality of any muscle food is obviously affected by the way it is processed. Why do we want to process food? At present, there are many modern reasons why foods are processed, e.g., adding value to a food, improving the visual appeal, convenience. However, traditionally, the single most important reason that we wish to pre-process food is to make them last longer without spoiling. Probably the oldest methods of achieving this goal are the salting of meat and fish, fermenting of milk, and pickling of vegetables.

Foods are made from natural materials, and like any living matter, will deteriorate in time. The deterioration of food, or food spoilage, is the natural way of recycling, restoring carbon, phosphorus, and nitrogenous matters to the good earth. However, putrefaction (spoilage) will modify the quality of foods resulting in poor appearance (discoloration), offensive smell, and inferior taste. Food spoilage can be caused by a number of factors, chiefly by biological factors, but also by chemical and physical factors. Consumption of spoiled foods can cause sickness and even death. There is no doubt none of us consider spoiled foods as having quality.

Selected examples will illustrate how food processing can affect the quality of a food product: Heat application. All of us know that overheating tender meat and chicken usually means toughness. The same is especially true for seafood. Heat removal or cold preservation. Freezing is a good example. Most of us are familiar with freezer-burn of meat, chicken, fish, shellfish, or other products left in the freezer over extended periods of time.

1. Evaporation and dehydration: Food drying has been popular since the beginning of time. Destruction of nutrients, especially vitamins, is one drawback to this method of preservation. Fermentation. In general, of meat, poultry, and fish products, fermented meat such as sausages is most popular. The quality of a sausage is to a large extent determined by the consumer, e.g., dry, sweet, salty, and pickled. Each method affects the quality in terms of nutrients, hardness, tenderness, and flavor.

2. New technology: There are numerous new technologies in food processing such as irradiation, microwaving, and ohmic heating. Each method affects the quality of a food in various ways. The finished product requires packaging. The obvious reason for packaging a food product, muscle foods or other, is to protect the food so it will not be exposed to the elements until it is ready to be prepared and consumed. The quality and shelf life of a food, especially a muscle food, depends very much on the way it is packaged.

Sensory Attributes and the Consumer

The sensory attributes of muscle foods are related to the senses of taste, smell, sight, feel, and sound. Of all the foods consumed, muscle foods have the lowest tolerance for complete sensorial acceptability. A muscle food is either acceptable or unacceptable with little in between. Predominately, the consumer visually assesses the color and surface texture of the muscle. The preparation technique of consumer choice is utilized, thereby altering the sensory attributes (usually completely).

The consumer cooks or prepares the muscle food as they prefer, changing the surface color, appearance, and texture. The internal altering of texture and flavor is a result of the preparation or cooking process as well. This will vary depending on the many methods applied. For instance, the muscle may be grilled, baked, broiled, or otherwise prepared, all with different fluctuating end results. Consumption of muscle foods is one of the most pleasurable eating experiences. The satiety value applied by the consumption of a muscle food is great when comparing the satisfying effect of foods in general. This is why the sensorial properties of muscle foods can be viewed as often more important than that of other foods.

Food Quality Management

Food quality management systems help to ensure that products are safe and fit for purpose. Food quality management uses tools, such as HACCP and auditing, to describe processes, analyse what could go wrong, put procedures in place to prevent things going wrong, check the systems are working, and identify actions to take if things do go wrong.

Food Fortification

Food fortification or enrichment is the process of adding micronutrients (essential trace elements and vitamins) to food. It can be carried out by food manufacturers, or by governments as a public health policy which aims to reduce the number of people with dietary deficiencies within a population. The predominant diet within a region can lack particular nutrients due to the local soil or from inherent deficiencies within the staple foods; addition of micronutrients to staples and condiments can prevent large-scale deficiency diseases in these cases.

As defined by the World Health Organization (WHO) and the Food and Agricultural Organization of the United Nations (FAO), fortification refers to “the practice of deliberately increasing the content of an essential micronutrient, ie. vitamins and minerals (including trace elements) in a food, so as to improve the nutritional quality of the food supply and to provide a public health benefit with minimal risk to health”, whereas enrichment is defined as “synonymous with fortification and refers to the addition of micronutrients to a food which are lost during processing”.

Food fortification has been identified as the second strategy of four by the WHO and FAO to begin decreasing the incidence of nutrient deficiencies at the global level. As outlined by the FAO, the most commonly fortified foods are cereals and cereal-based products; milk and dairy products; fats and oils; accessory food items; tea and other beverages; and infant formulas. Undernutrition and nutrient deficiency is estimated globally to cause the deaths of between 3 and 5 million people per year.

Types

Main methods of food fortification:

1. Commercial and industrial fortification (wheat flour, corn meal, cooking oils).
2. Biofortification (breeding crops to increase their nutritional value, which can include both conventional selective breeding, and genetic engineering).
3. Home fortification (example: vitamin D drops).

Rationale

The WHO and FAO, among many other nationally recognized organizations, have recognized that there are over 2 billion people worldwide who suffer from a variety of micronutrient deficiencies. In 1992, 159 countries pledged at the FAO/WHO International Conference on Nutrition to make efforts to help combat these issues of micronutrient deficiencies, highlighting the importance of decreasing the number of those with iodine, vitamin A, and iron deficiencies. A significant statistic that led to these efforts was the discovery that approximately 1 in 3 people worldwide were at risk for either an iodine, vitamin A, or iron deficiency. Although it is recognized that food fortification alone will not combat this deficiency, it is a step towards reducing the prevalence of these deficiencies and their associated health conditions.

In Canada, the Food and Drug Regulations have outlined specific criterion which justifies food fortification:

1. To replace nutrients which were lost during manufacturing of the product (e.g. the manufacturing of flour).
2. To act as a public health intervention.
3. To ensure the nutritional equivalence of substitute foods (e.g. to make butter and margarine similar in content, soy milk and cow's milk, etc.).
4. To ensure the appropriate vitamin and mineral nutrient composition of foods for special dietary purposes (e.g., gluten-free products, low sodium, or any other products specifically designed for special dietary requirements from an individual).

There are also several advantages to approaching nutrient deficiencies among populations via food fortification as opposed to other methods. These may include, but are not limited to: treating a population without specific dietary interventions therefore not requiring a change in dietary patterns, continuous delivery of the nutrient, does not require individual compliance, and potential to maintain nutrient stores more efficiently if consumed on a regular basis.

Fortified food.

Several organizations such as the WHO, FAO, Health Canada, and Nestlé Research acknowledge that there are limitations to food fortification. Fortification of nutrients in foods may deliver excessive amounts of nutrients to some individuals, with consequent side effects. One example is fluoride, which can cause irreversible staining to the teeth. Another example is iron, as fortification intended to benefit women may result in too much iron consumption by men.

The WHO states that limitations to food fortification may include human rights issues indicating that consumers have the right to choose if they want fortified products or not, the potential for insufficient demand of the fortified product, increased production costs leading to increased retail costs, the potential that the fortified products will still not be a solution to nutrient deficiencies amongst low income populations who may not be able to afford the new product, and children who may not be able to consume adequate amounts thereof.

In addition to criticism of government-mandated fortification, food companies have been criticized for indiscriminate enrichment of foods for marketing purposes. Food safety worries led to legislation in Denmark in 2004 restricting foods fortified with extra vitamins or minerals. Products banned include: Rice Crispies, Shreddies, Horlicks, Ovaltine and Marmite.

Limited Absorption

One factor that limits the benefits of food fortification is that isolated nutrients added back into a processed food that has had many of its nutrients removed, does not always result in the added nutrients being as bioavailable as they would be in the original, whole food. An example is skim milk that has had the fat removed, and then had vitamin A and vitamin D added back. Vitamins A and D are both fat-soluble and non-water-soluble, so a person consuming skim milk in the absence of fats may not be able to absorb as much of these vitamins as one would be able to absorb from drinking whole milk. On the other hand, the nutrient added as a fortificant may have a higher bioavailability than from foods, which is the case with folic acid used to increase folate intakes.

Phytochemicals such as phytic acid in cereal grains can also impact nutrient absorption, limiting the bioavailability of intrinsic and additional nutrients, and reducing the effectiveness of fortification programs.

Different Forms of Micronutrients

There is a concern that micronutrients are legally defined in such a way that does not distinguish between different forms, and that fortified foods often have nutrients in a balance that would not occur naturally. For example, in the U.S., food is fortified with folic acid, which is one of the many naturally-occurring forms of folate, and which only contributes a minor amount to the folates occurring in natural foods. In many cases, such as with folate, it is an open question of whether or not there are any benefits or risks to consuming folic acid in this form.

In many cases, the micronutrients added to foods in fortification are synthetic.

In some cases, certain forms of micronutrients can be actively toxic in a sufficiently high dose, even if other forms are safe at the same or much higher doses. There are examples of such toxicity in both synthetic and naturally-occurring forms of vitamins. Retinol, the active form of Vitamin A, is toxic in a much lower dose than other forms, such as beta carotene. Menadione, a phased-out synthetic form of Vitamin K, is also known to be toxic.

Examples of Fortification in Foods

Many foods and beverages worldwide have been fortified, whether a voluntary action by the product developers or by law. Although some may view these additions as strategic marketing schemes to sell their product, there is a lot of work that must go into a product before simply fortifying it. In order to fortify a product, it must first be proven that the addition of this vitamin or mineral is beneficial to health, safe, and an effective method of delivery. The addition must also abide by all food and labeling regulations and support nutritional rationale. From a food developer's point of view, they also need to consider the costs associated with this new product and whether or not there will be a market to support the change.

The Food Fortification Initiative lists all countries in the world that conduct fortification programs, and within each country, what nutrients are added to which foods, and whether those programs are voluntary or mandatory. Vitamin fortification programs exist in one or more countries for folate, niacin, riboflavin, thiamin, vitamin A, vitamin B_6, vitamin B_{12}, vitamin D and vitamin E. Mineral fortification programs include calcium, fluoride, iodine, iron, selenium and zinc. As of December 21, 2018, 81 countries required food fortification with one or more vitamins. The most commonly fortified vitamin – as used in 62 countries – is folate; the most commonly fortified food is wheat flour. Examples of foods and beverages that have been fortified.

Iodised Salt

Iodine deficiency disorder (IDD) is the single greatest cause of preventable mental retardation. Severe deficiencies cause cretinism, stillbirth and miscarriage. But even mild deficiency can significantly affect the learning ability of populations. Today over 1 billion people in the world suffer from iodine deficiency, and 38 million babies born every year are not protected from brain damage due to IDD.

Iodised salt has been used in the United States since before World War II. It was discovered in 1821 that goiters could be treated by the use of iodized salts. However, it was not until 1916 that the use of iodized salts could be tested in a research trial as a preventative measure against goiters. By 1924, it became readily available in the US. Currently in Canada and the US, the RDA for iodine is as low as 90 μg/day for children (4–8 years) and as high as 290 μg/day for breast-feeding mothers.

Diseases that are associated with an iodine deficiency include: mental retardation, hypothyroidism, and goiter. There is also a risk of various other growth and developmental abnormalities.

Folate

Folate (as a fortification ingredient, folic acid) functions in reducing blood homocysteine levels, forming red blood cells, proper growth and division of cells and preventing neural tube defects (NTDs). In many industrialized countries, the addition of folic acid to flour has prevented a significant number of NTDs in infants. Two common types of NTDs, spina bifida and anencephaly, affect approximately 2500-3000 infants born in the US annually. Research trials have shown the ability to reduce the incidence of NTDs by supplementing pregnant mothers with folic acid by 72%.

Niacin

Niacin has been added to bread in the US since 1938 (when voluntary addition started), a program which substantially reduced the incidence of pellagra. Pellagra was seen amongst poor families who used corn as their main dietary staple. Although corn itself does contain niacin, it is not a bioavailable form unless it undergoes nixtamalization (treatment with alkali, traditional in Native American cultures) and therefore was not contributing to the overall intake of niacin.

Diseases associated with niacin deficiency include: Pellagra which consisted of signs and symptoms called the three D's-"Dermatitis, dementia, and diarrhea." Others may include vascular or gastrointestinal diseases. Common diseases which present a high frequency of niacin deficiency: alcoholism, anorexia nervosa, HIV infection, gastrectomy, malabsorptive disorders, certain cancers and their associated treatments.

Vitamin D

Since Vitamin D is a fat-soluble vitamin, it cannot be added to a wide variety of foods. Foods that it is commonly added to are margarine, vegetable oils and dairy products. During the late 1800s, after the discovery of curing conditions of scurvy and beriberi had occurred, researchers were aiming to see if the disease, later known as rickets, could also be cured by food. Their results showed that sunlight exposure and cod liver oil were the cure. It was not until the 1930s that vitamin D was actually linked to curing rickets. This discovery led to the fortification of common foods such as milk, margarine, and breakfast cereals. This took the astonishing statistics of approximately 80–90% of children showing varying degrees of bone deformations due to vitamin D deficiency to being a very rare condition.

Diseases associated with a vitamin D deficiency include rickets, osteoporosis, and certain types of cancer (breast, prostate, colon and ovaries). It has also been associated with increased risks for fractures, heart disease, type 2 diabetes, autoimmune and infectious diseases, asthma and other wheezing disorders, myocardial infarction, hypertension, congestive heart failure, and peripheral vascular disease.

Fluoride

Although fluoride is not considered an essential mineral, it is useful in prevention of tooth decay and maintaining adequate dental health. In the mid-1900s it was discovered that towns with a high level of fluoride in their water supply was causing the residents' teeth to have both brown spotting and a strange resistance to dental caries. This led to the fortification of water supplies with fluoride in safe amounts (or reduction of naturally-occurring levels) to retain the properties of resistance to dental caries but avoid the staining cause by fluorosis (a condition caused by excessive fluoride intake). The tolerable upper intake level (UL) set for fluoride ranges from 0.7 mg/day for infants aged 0–6 months and 10 mg/day for adults over the age of 19.

Fat Substitute

A fat substitute is a food product with the same functions, stability, physical, and chemical characteristics as regular fat, with fewer Calories per gram than fat. They are utilized in the production of low fat and low calorie foods.

Fat is present in most foods. It provides a unique texture, flavor, and aroma to the food it is found in. While fat is essential to life, it can be detrimental to health when consumed in excess of physiological requirements. High fat diets increase risk of heart disease, weight gain, and some cancers. High blood cholesterol is more prevalent in those that consume diets high in saturated fats, and it increases risk for coronary heart disease in those individuals. The use of fat substitutes in food products allows for maintenance of the food's original quality characteristics without the associated risks of fat consumption. In the absence of energy-dense fat molecules, products utilizing fat substitutes are generally lower in calories than their full-fat counterparts.

Applications

Fat substitutes can be divided into four categories based on the food component from which they are derived.

Category	Type and example	Function
Carbohydrate-based	• Cellulose (Vivapur) • Dextrins, modified starches (Stellar) • Fruit-based fibre (WonderSlim) • Grain-based fibre (Betatrim) • Hydrocolloid gums • Maltodextrin (Maltrin) • Pectin (Grinsted)	Binder, body, bulk, flavor, moisture retention, mouth feel
Protein-based	• Microparticulate protein (Simplesse) • Modified whey protein concentrate (Dairy-Lo)	Mouth feel, water-binding, reduce syneresis
Fat-based	• Altered triglycerides (Caprenin) • Sucrose polyesters (Olestra) • Esterified propoxylated glycerol (EPG)	Emulsion, mouth feel
Combination	• Carbohydrate and protein (Mimix) • Carbohydrate and fat (Optamax)	Flavour, texture, mouth feel, water retention

Potential Benefits

Consumption of fat substitutes can assist in lowering total overall fat and calorie intake from foods. This has positive implications for those looking to reduce either one of these, especially when in a disease state associated with high fat diets. While fat substitution alone can reduce the percentage of kilocalories ingested from dietary fat, it may not reduce an individual's total energy intake (in terms of kilocalories) unless the rest of the diet is of high quality and low energy density.

Safety

Few concerns have been raised about the safety of fat substitutes. Carrageenan, olestra, and polydextrose have been approved by the U.S. Food and Drug Administration (FDA) for use as food additives, a title which requires both intensive testing over a wide demographic and the ability to meet strict, pre-determined, FDA criteria. Other products, such as guar gum and maltodextrose, are "Generally Recognized as Safe (GRAS)" by the FDA; this is also based on scientific testing and long-term consumption by a variety of consumer demographics. With excessive use, polydextrose can have a laxative effect, and olestra may cause loss of fat-soluble vitamins in the form of fatty stools and is liquid at body temperature. Esterified propoxylated glycerol (EPG), which is a solid at body temperature, achieved GRAS status for confectionery uses in November, 2015. EPG's GRAS status expanded to use at levels up to 38 percent by weight in baked goods and baking mixes, frozen dairy desserts and mixes, grain products and pasta, gravies and sauces, nuts and nut products, and soft candy. At this time, there is little supporting evidence to accompany claims that these, or other fat substitutes, are hazardous; however, more long-term research is needed.

Food Grading

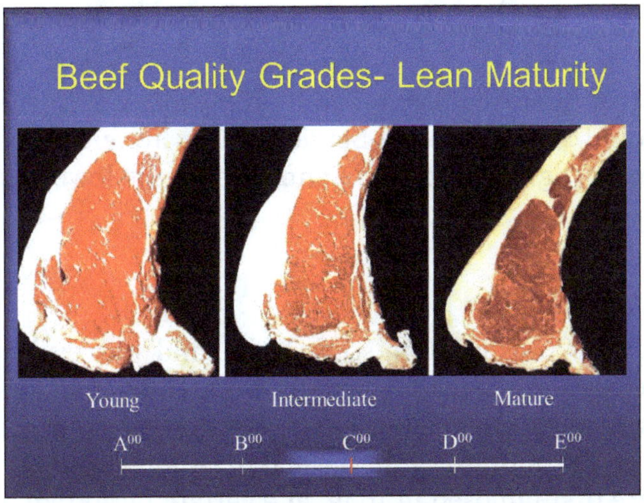

Optical sorting achieves non-destructive, 100 percent inspection in-line at full production volumes.

Food grading involves the inspection, assessment and sorting of various foods regarding quality, freshness, legal conformity and market value. Food grading often occurs by hand, in which foods are assessed and sorted. Machinery is also used to grade foods, and may involve sorting products by size, shape and quality. For example, machinery can be used to remove spoiled food from fresh product.

By Food Type

Beef

Inspected beef carcasses tagged by the USDA.

Beef grading in the United States is performed by the United States Department of Agriculture's (USDA) Agricultural and Marketing Service. There are eight beef quality grades, with U.S. Prime being the highest grade and U.S. Canner being the lowest grade. Beef grading is a complex process.

Beer

In beer grading, the letter "X" is used on some beers, and was traditionally a mark of beer strength, with the more Xs the greater the strength. Some sources suggest that the origin of the mark was in the breweries of medieval monasteries Another plausible explanation is contained in a treatise entitled "The Art of Brewing" published in London in 1829. It says; "The duties on ale and beer, which were first imposed in 1643, at a certain period, in distinguishing between small beer and strong, all ale or beer, sold at or above ten shillings per barrel, was reckoned to be strong and was, therefore, subjected to a higher duty. The cask which contained this strong beer was then first marked with an X signifying ten; and hence the present quack-like denominations of XX (double X) and XXX (treble X) on the casks and accounts of the strong-ale brewers".

In mid-19th century England, the use of "X" and other letters had evolved into a standardised grading system for the strength of beer. Today, it is used as a trade mark by a number of brewers in the United Kingdom, the Commonwealth and the United States.

European Bitterness Units scale, often abbreviated as EBU, is a scale for measuring the perceived bitterness of beer, with lower values being generally "less bitter" and higher values "more bitter". The scale and method are defined by the European Brewery Convention, and the numerical value should be the same as of the International Bitterness Units scale (IBU), defined in co-operation with the American Society of Brewing Chemists. However, the exact process of determining EBU and IBU values differs slightly, which may in theory result with slightly smaller values for EBU than IBU.

The International Bittering Units scale, or simply IBU scale, provides a measure of the bitterness of beer, which is provided by the hops used during brewing. Bittering units are measured through the use of a spectrophotometer and solvent extraction.

Butter

Coconut Milk

Several grades of coconut milk exist: from thick at 20-22% fat to thin at 5-7% fat level.

Coffee

Dark-roasted coffee beans.

Coffee growers, traders, and roasters grade beans based on size, color, and a variety of other characteristics. Coffees of exceptional quality are traded as "specialty coffees" and fetch a higher price in the international market.

After the roast, Coffee grading involves assessment of roasted coffee seed colorization and then labeling as light, medium light, medium, medium dark, dark, or very dark. A more accurate method of discerning the degree of roast involves measuring the reflected light from roasted seeds illuminated with a light source in the near infrared spectrum. This elaborate light meter uses a process known as spectroscopy to return a number that consistently indicates the roasted coffee's relative degree of roast or flavor development.

Eggs

In the United States, egg grading is performed by the USDA, and is based upon the interior quality of the egg and the appearance and condition of the egg shell. Eggs of any quality grade may differ in weight (size). Egg grading is performed by candling, which involves observing the interior of eggs by placing them in front of a bright light.

Guar Gum

Guar gum grading involves analysis for coloration, viscosity, moisture, granulation, protein content and insolubles ash.

Honey

Honey grading in the United States is performed voluntarily based upon USDA standards (USDA does offer inspection and grading "as on-line (in-plant) or lot inspection upon application, on a fee-for-service basis.") . Honey is graded based upon a number of factors, including water content, flavor and aroma, absence of defects and clarity. Honey is also classified by color though it is not a factor in the grading scale. U.S. honey grade scales are Grade A, Grade B, Grade C and Grade substandard.

Lobster

A cooked lobster.

In the U.S., lobster grading involves denoting lobsters as new-shell, hard-shell or old-shell, and because lobsters which have recently shed their shells are the most delicate, there is an inverse relationship between the price of American lobster and its flavour. New-shell lobsters have paper-thin shells and a worse meat-to-shell ratio, but the meat is very sweet. However, the lobsters are so delicate that even transport to Boston almost kills them, making the market for new-shell lobsters strictly local to the fishing towns where they are offloaded. Hard-shell lobsters with firm shells, but with less sweet meat, can survive shipping to Boston, New York and even Los Angeles, so they command a higher price than new-shell lobsters. Meanwhile, old-shell lobsters, which have not shed since the previous season and have a coarser flavour, can be air-shipped anywhere in the world and arrive alive, making them the most expensive. One seafood guide notes that an eight-dollar lobster dinner at a restaurant overlooking fishing piers in Maine is consistently delicious, while "the eighty-dollar lobster in a three-star Paris restaurant is apt to be as much about presentation as flavor".

Maple Syrup

Following an effort from the International Maple Syrup Institute (IMSI) and many maple syrup producer associations, both Canada and the United States have altered their laws regarding the classification of maple syrup to be uniform. Whereas in the past each state or province had their own laws on the classification of maple syrup, now those laws define a unified grading system. This had been a work in progress for several years, and most of the finalization of the new grading system was made in 2014. The Canadian Food Inspection Agency announced in the Canada Gazette on 28 June 2014 that rules for the sale of maple syrup would be amended to include new descriptors, at the request of the IMSI.

As of December 31, 2014, the Canadian Food Inspection Agency (CFIA) and as of March 2, 2015, the United States Department of Agriculture (USDA) Agricultural Marketing Service (AMS) issued revised standards on the classification of maple syrup as follows:

- Grade A:
    - Golden Color and Delicate Taste,
    - Amber Color and Rich Taste,

    - Dark Color and Robust Taste,
    - Very Dark Color and Strong Taste.
- Processing Grade,
- Substandard.

As long as maple syrup does not have an off-flavor and is of a uniform color and clean and free from turbidity and sediment, it can be labelled as one of the A grades. If it exhibits any of these problems, it does not meet Grade A requirements and must be labelled as Processing Grade maple syrup and may not be sold in containers smaller than 5 gallons. If maple syrup does not meet the requirements of Processing Grade maple syrup (including a fairly characteristic maple taste), it is classified as Substandard. As of February 2015, this grading system has been accepted and made law by most maple-producing states and provinces, other than Ontario, Quebec, and Ohio. Vermont, in an effort to "jump-start" the new grading regulations, adopted the new grading system as of January 1, 2014, after the grade changes passed the Senate and House in 2013. Maine passed a bill to take effect as soon as both Canada and the United States adopted the new grades. They are allowing a one-year grace period. In New York, the new grade changes became law on January 1, 2015, with a one-year grace period. New Hampshire did not require legislative approval and so the new grade laws became effective as of December 16, 2014, and producer compliance was required as of January 1, 2016.

Golden and Amber grades typically have a milder flavor than Dark and Very dark, which are both dark and have an intense maple flavor. The darker grades of syrup are used primarily for cooking and baking, although some specialty dark syrups are produced for table use. Syrup harvested earlier in the season tends to yield a lighter color. With the new grading system, the classification of maple syrup depends ultimately on its internal transmittance at 560 nm wavelength through a 10 mm sample. Golden has to have more than 75 percent transmittance, Amber has to have 50.0 to 74.9 percent transmittance, Dark has to have 25.0 to 49.9 percent transmittance, and Very Dark is any product less than 25.0 percent transmittance.

Old Grading System

Old US maple syrup grades, left to right: Grade A Light Amber ("Fancy"), Grade A Medium Amber, Grade A Dark Amber, Grade B.

In Canada, maple syrup was classified prior to December 31, 2014, by the Canadian Food Inspection Agency (CFIA) as one of three grades, each with several color classes: Canada No. 1, including

Extra Light, Light, and Medium; No. 2 Amber; and No. 3 Dark or any other ungraded category. Producers in Ontario or Québec may have followed either federal or provincial grading guidelines. Québec's and Ontario's guidelines differed slightly from the federal: there were two "number" categories in Québec (Number 1, with four color classes, and 2, with five color classes). As in Québec, Ontario's producers had two "number" grades: 1, with three color classes; and 2, with one color class, which was typically referred to as "Ontario Amber" when produced and sold in that province only. A typical year's yield for a maple syrup producer will be about 25 to 30 percent of each of the #1 colors, 10 percent #2 Amber, and 2 percent #3 Dark. Producers in Quebec and Ontario may follow either federal or provincial grading guidelines, which differ slightly.

The United States used (some states still do, as they await state regulation) different grading standards. Maple syrup was divided into two major grades: Grade A and Grade B. Grade A was further divided into three subgrades: Light Amber (sometimes known as Fancy), Medium Amber, and Dark Amber. The Vermont Agency of Agriculture Food and Markets used a similar grading system of color, and is roughly equivalent, especially for lighter syrups, but using letters: "AA", "A", etc. The Vermont grading system differed from the US system in maintaining a slightly higher standard of product density (measured on the Baumé scale). New Hampshire maintained a similar standard, but not a separate state grading scale. The Vermont-graded product had 0.9 percent more sugar and less water in its composition than US-graded. One grade of syrup not for table use, called commercial or Grade C, was also produced under the Vermont system. Vermont inspectors enforce strict syrup grading regulations, and can fine producers up to US$1000 for labelling syrup incorrectly.

Milk

In the United States, there are two grades of milk, with Grade A primarily used for direct sales and consumption in stores, and Grade B used for indirect consumption, such as in cheese making or other processing.

The two grades are defined in the Wisconsin Administrative Code. Grade B generally refers to milk that is cooled in milk cans, which are immersed in a bath of cold flowing water that typically is drawn up from an underground water well rather than using mechanical refrigeration.

Oranges

Florida navel oranges.

The USDA has established the following grades for Florida oranges, which primarily apply to oranges sold as fresh fruit: US Fancy, US No. 1 Bright, US No. 1, US No. 1 Golden, US No. 1 Bronze,

US No. 1 Russet, US No. 2 Bright, US No. 2, US No. 2 Russet, and US No. 3. The general characteristics graded are color (both hue and uniformity), firmness, maturity, varietal characteristics, texture, and shape. Fancy, the highest grade, requires the highest grade of color and an absence of blemishes, while the terms Bright, Golden, Bronze, and Russet concern solely discoloration.

Peas

Peas.

Pea grading involves sorting peas by size, in which smallest peas are graded as the highest quality for their tenderness. Brines may be used, in which peas are floated in them, from which their density can be determined.

Potatoes

In the U.S., potato grading for Idaho potatoes is performed in which No. 1 potatoes are the highest quality and No. 2 are rated as lower in quality due to their appearance (e.g. blemishes or bruises, pointy ends). Density assessment can be performed by floating them in brines. High density potatoes are desirable in the production of dehydrated mashed potatoes, potato crisps and french fries.

Rice

The main criteria used by many countries and millers in rice grading are degree of milling, appearance (color), damaged (broken) and percentage of chalky kernels. In the United States rice is marketed according to three main properties size, color and condition (kernels damage), these properties are directly related to quality, milling percentage and other processing conditions. All properties are considered important in grading. For instance, chalky kernels are not desirable because they give lower milling yields after processing and easily break during handling.

Spices

Cinnamon

In Sri Lanka, cinnamon grading is performed by dividing cinnamon quills into four groups, which are then further divided into specific grades.

Vanilla

People grading vanilla beans in Sambava, Madagascar.

Several vanilla fruit grading systems are in use. Each country which produces vanilla has its own grading system, and individual vendors, in turn, sometimes use their own criteria for describing the quality of the fruits they offer for sale.

Tea

In the western black tea industry, tea leaf grading is the process of evaluating products based on the quality and condition of the tea leaves themselves. The highest grades are referred to as "orange pekoe", and the lowest as "fannings" or "dust". This grading system is based upon the size of processed and dried black tea leaves. Despite a purported Chinese origin, these grading terms are typically used for teas from Sri Lanka, India and countries other than China; they are not generally known within Chinese-speaking countries.

Black Tea

Black tea grading is usually based upon one of four scales of quality. Whole-leaf teas are the highest quality, followed by broken leaves, fannings, and dusts. Whole-leaf teas are produced with little or no alteration to the tea leaf. This results in a finished product with a coarser texture than that of bagged teas. Whole-leaf teas are widely considered the most valuable, especially if they contain leaf tips. Broken leaves are commonly sold as medium-grade loose teas. Smaller broken varieties may be included in tea bags.

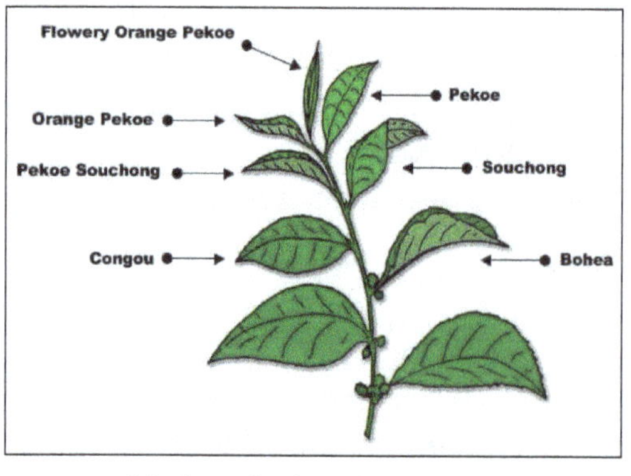

A black tea leaf grading diagram.

Fresh tea leaves of different sizes.

Rooibos

Rooibos grades are largely related to the proportion of "needle" or leaf to stem content in the mix. A higher leaf content will result in a darker liquor, richer flavour and less "dusty" aftertaste. The high-grade rooibos is exported and does not reach local markets, with major consumers being in the EU, particularly Germany, where it is used in creating flavoured blends for loose-leaf tea markets. In development within South Africa are a small number of specialty tea companies producing similar blends.

Organic green rooibos tea leaves.

Food Additives and Preservatives

Carrageenan

There are two basic grades of carrageenan, refined carrageenan (RC) and semi-refined carrageenan (SRC). In the United States, RC and SRC are both labeled as carrageenan. In the European Union, RC is designated by the E number E-407, and SRC is E-407a. RC has a 2% maximum for acid-insoluble material and is produced through an alcohol precipitation process or potassium chloride gel press process. SRC contains a much higher level of cellulosic content and is produced in a less complex process. Indonesia, the Philippines, and Chile are three main sources of raw material and extracted carrageenan.

Lye

Lye is used to cure foods such as lutefisk, olives (making them less bitter), canned mandarin oranges, hominy, lye rolls, century eggs, and pretzels. It is also used as a tenderizer in the crust of baked Cantonese mooncakes, and in lye-water "zongzi" (glutinous rice dumplings wrapped in bamboo leaves), chewy southern Chinese noodles popular in Hong Kong and southern China, and Japanese ramen noodles. In the United States, food-grade lye must meet the requirements outlined in the Food Chemicals Codex (FCC), as prescribed by the U.S. Food and Drug Administration (FDA). Lower grades of lye are commonly used as drain or oven cleaner. Such grades should not be used for food preparation, as they may contain impurities harmful to human health.

Sodium Bisulphate

Sodium bisulphate is used as a food additive to leaven cake mixes (make them rise) as well as being used in meat and poultry processing and most recently in browning prevention of fresh-cut produce. The food-grade product meets the requirements set out in the Food Chemicals Codex.

It is denoted by E number E514ii in the EU and is approved for use in Australia and New Zealand where it is listed as additive 514. Food-grade sodium bisulfate is used in a variety of food products, including beverages, dressings, sauces, and fillings.

By Country

- In India, AGMARK is a certification mark employed on agricultural products, assuring that they conform to a set of standards approved by the Directorate of Marketing and Inspection, an agency of the Government of India.

Food Sampling

A procedure used to draw inferences about a lot (population) from results obtained from a sample.

Sampling involves the selection of a certain portion, number of container and product units from a particular lot of the same food. It must be as representative as possible of the whole consignment or from lot.

An act of obtaining a sample is called sampling, which can be done by a person or automatically. Samples of material can be taken or provided for testing, analysis, inspection, investigation, demonstration, or trial use.

Samples are usually collected from a lot of food for random surveillance, collection of data for a specific purpose, or monitoring/and to determine whether the food is unsatisfactory for any reason.

Importance of Sample Collection

The reliability of analytical data thus obtained depends on several factors, sampling being the major factor. Current analytical methods require only few grams of food sample to analyze. Thus, it is necessary that a sample be as representative of the population as possible.

Food samples of biological origin (liquid or solid) have been divided generally into the five categories described in Table This coarse division is important when considering the choice of isolation technique, extraction solvent, and sample clean-up method during an analytical procedure.

Moisture content is an important consideration during sampling procedures, in part because it affects the extent of sample heterogeneity. Virtually all foods are heterogeneous, and the analyst should be familiar with their variability in composition and structure. In general, fresh foods of plant origin are more variable in composition than fresh foods of animal origin. The analyst should be also aware of the postmortem or postharvest physiological changes that can occur after a fresh food is sampled and which can affect sample heterogeneity. A combination of cold storage and chemical preservation may be required to maintain sample integrity in the event of prolonged storage.

Table: Classification of food samples according to their content.

Sample	Character	Typical analytes
Milk	Aqueous, proteins, lipids	Veterinary drugs, toxic elements, pesticides, industrial contaminants
Eggs	High lipids and albumin content	Veterinary drugs , pesticides, industrial contaminants
Other samples of animal origin (liver, fat)	Various fat, proteins, or water	Drugs, pesticides, industrial contaminants
Plant materials (fruits, vegetables)	Various water, plant pigments, proteins, lipids, essential oil, waxes	toxic elements, pesticides, industrial contaminants
Food (meat, milk, cereals, wines, juices, plants oils, sugar	Various fat, oils, lipids, proteins, sugar, starch, water, or pigments	pesticides, industrial contaminants, synthetic colorants, additives, synthetic sweeteners, antioxidants

Although the chemical and physical properties of foods are inherently variable, even between samples that originate from the same breed or strain, the variability in composition of a single food sample can be minimized with proper sampling and sample pretreatment techniques. Two approaches can be used for sampling a food mass that is larger than the amount required for analysis in the laboratory. Many minute increments of a solid material can be collected and blended to represent the entire foodstuff, or a quantity of material that is large enough to be compositionally representative of the whole can be collected and then reduced to a fine mixture before being sub-sampled. The first approach is usually avoided, since it is difficult to obtain a statistically representative sample and the sampling time can also be very long. The latter approach is more practical, accurate, and reproducible.

Since virtually no food material can be analyzed in its entirety, careful sampling techniques are required to obtain representative, laboratory-sized primary samples, in addition to subsequent subsamples, or secondary samples. The amount of subsample required for an analytical procedure usually varies from a fraction of a gram to several grams.

Sample Size

The required sample size is defined in part by the nature of the target compound, that is, to what extent the analyte is retained in the matrix. Xenobiotics are generally present at trace levels. A sufficiently large amount of sample must be collected and analyzed in order to be able to measure minute quantities of the compound of interest and to satisfy the method's limit of detection. Conversely, relatively small samples may be collected for the macro analysis of gross food components, i.e., to measure crude fat, crude protein, crude fiber, or ash. Although proximate analysis of these food components is sometimes sufficient, more exact analyses are usually required.

The sample size is also dependent on the relationship that exists between the mass required to adequately represent a sample and the characteristics of that sample. If a foodstuff consists of some mixture of different-sized particles, enough sample mass needs to be collected in order to adequately represent all of the particles. Because large particles are more difficult to represent than smaller ones, a mass that is large enough to represent the larger particles will also be representative of the smaller ones. The segregation of finer, denser particles to the bottom of the sample container must be recognized during the sampling process to ensure that all particles are represented and to avoid large sampling errors.

Sampling Steps

Sample Collection

- Containers
- Sampling devices
- Sampling procedures
- Sample labeling

Techniques

Food lots are sampled in either a manual or continuous manner in order to obtain a representative specimen. Containers holding loose foodstuffs can be sampled manually with devices that trap the material in a compartment such as a probe or tube. Slots or openings placed at intervals in the tube allow for simultaneous sampling at different depths of the product. When employing this technique, however, the analyst must consider the segregation effect and ensure that all particle sizes are accessible. The foodstuff may ultimately need to be removed from the sample-container and poured onto a flat surface. The amount of material may then be reduced with a coning-and-quartering method, and a subsample collected in multiple random increments. No particle size should be excluded during the sampling process. Since food components or contaminants that collect in certain-sized particles might be omitted from the final analysis, thereby resulting in an increase in sampling error.

Large mixtures may also be reduced with a riffle cutter, which is a box-like device that has equally spaced dividers to divide the sample stream. The sample may be further cut or quartered by passing it through successive riffles. Other proportional dividers are available for reducing a sample, such as the straightline sampler and the spinning riffle sample divider.

Uniformly solid or liquid products are perhaps the most straightforward to sample. Drill-type devices are used to obtain a core from solid products such as cheese or frozen foods. Liquid samples are thoroughly mixed before a subsample is removed with a syringe-type sampler or by submerging a container under the liquid's surface (a so-called "grab" sample). For obvious reasons, 'many complex foods such as vegetables, fruit, or animal tissues may require blending prior to being sampled.

Throughout the sample preparation procedure, it is essential for the analyst to recognize the necessity of utilizing methods that satisfy statistical sampling and analysis requirements. The inherent variability in the composition of raw materials, basic ingredients, and processed foods requires the use of statistical methods for obtaining representative and replicate samples, and for estimating the error involved in sampling.

Food Pre-treatment

Removal of Extraneous Matter

Before sample blending is done, it is often necessary to wash, remove, or drain irrelevant extraneous matter. Soil or sand that adheres to fresh fruit or vegetables can be removed by washing or wiping the surface of the produce; however, excessive washing should be avoided to prevent the leaching of soluble solids.

Depending on the objective of the analysis, fresh produce may be separated into the core and the outer and inner tissues. Shells are usually separated from nut kernels and pits from stone fruits. Large fish are cleaned, scaled, and eviscerated, while small fish can be blended whole. Shellfish are shucked, eggs are broken to isolate the liquid interior, and meat is removed as completely as possible from bone. Canned fruit and vegetable products may be drained through screens if it is not necessary to analyze the composite sample.

Sample Reduction

Once a food sample has been collected using the sampling techniques a suitable method is required to make the material less heterogeneous. Various approaches may be utilized for reducing the particle weight and size in a primary sample, so that smaller subsamples can be taken for a representative analysis of the whole. Finely divided materials also dissolve faster and are easier to extract because of their greater surface area.

Methods for reducing solid or semi-solid foods include mechanical grinding, mixing, rolling, agitating, stirring, chopping, crushing, macerating, mincing, pressing, pulverizing, or any other reasonable means of cornminuting the sample.

Sample reduction can also be achieved with a Wiley or ball mill, mortar and pestle, mechanical high-speed beaters or blenders (for soft or wet foods), and meat grinders.

Sample handling:

- Transportation.
- Reception.

Sample analysis:

- Withdrawing analytical units.
- Homogenization of analytical units/Analysis using appropriate methods and instruments.

Types of Sampling

Bulk Sampling

It involves the selection of a sample from a lot of material that does not consist of discrete, identifiable or constant units. Sampling may be performed in static or dynamic situations. Bulk sampling poses special problems requiring certain decisions to be made: the number of increments to be taken, the size of the increments, from where in the pile or stream they should be drawn, the sampling device to be used, and how to reduce the increments taken to a reasonable size of ample for delivery in the laboratory.

Acceptance Sampling

It differs from the bulk sampling and involves the application of predetermined plan to decide whether a lot of goods meet defined criteria for acceptance. The risks of accepting –bad or rejecting “good” lots are stated in conjunction with one or more parameters. Statistical plans can be designed to regulate the probabilities of rejecting good lots or accepting bad lots.

Sampling Plan

The particular choice of sampling procedure to determine the minimum number of food units that will provide a high degree of certainty about the quality of a food lot.

Sample Units (n)

- Large enough to represent the population.
- Small enough to be economically feasible.

Sample Characteristics

The material may be solid, liquid, gas, material of some intermediate characteristics such as gel, tissue, organisms, or a combination of these. Even if a material sample is not countable as individual items, the quantity of the sample may still be describable in terms of its volume, mass, size, or other such dimensions. A solid sample can come in one or a few discrete pieces, or can be fragmented, granular, or powdered.

Solid Food Sampling

Samplers for General usage

- Sampling from bulk: Use appropriate apparatus for obtaining increments from static bulk (example, hand- held spears, mechanical or air-assisted apparatus).
- Sampling from bags: Use sack type spears.
- Mixing and dividing: Use shovels and dividing apparatus or automatic random dividing apparatus.

Sampling from Silos, Bins or Warehouses

Increments shall be taken throughout the whole depth of the lot. Suitable instruments must be used to achieve this requirement. If the depth of the lot does not permit use of this method, sampling should be carried out from the flowing cereal in accordance with ISO 6644.

Take the square root of the tonnage in the static bulk. Divide by two and round up to the next whole number. This is the minimum number of increments that is to be obtained.

Example: Number of increments for bulk grain of more than 500 t.

Tonnage	500	1000	2000	4000	6000	8000	10000
Square root	22.4	31.6	44.7	63.2	77.4	89.4	100
Number of increments	12	16	23	32	39	45	50

Sampling from bags: Unless otherwise specified in the contract or unless the practice at the port or elsewhere requires otherwise, increments shall be taken from different part of a bag (for eg. Top, middle, bottom) by means of a sack/ bag spear from the number of bags specified in table.

Table: Sampling Scheme for consignment of more than 100 bags.

Number of bags in consignment	Number of bags to be sampled
Up to 10	Each bag
10 to 100	10, taken at random
More than 100	Square root (approx.) of total number

The consignment shall be dividing into (n-1) groups containing n or (n-1) bags: the remaining bags constitute a group.

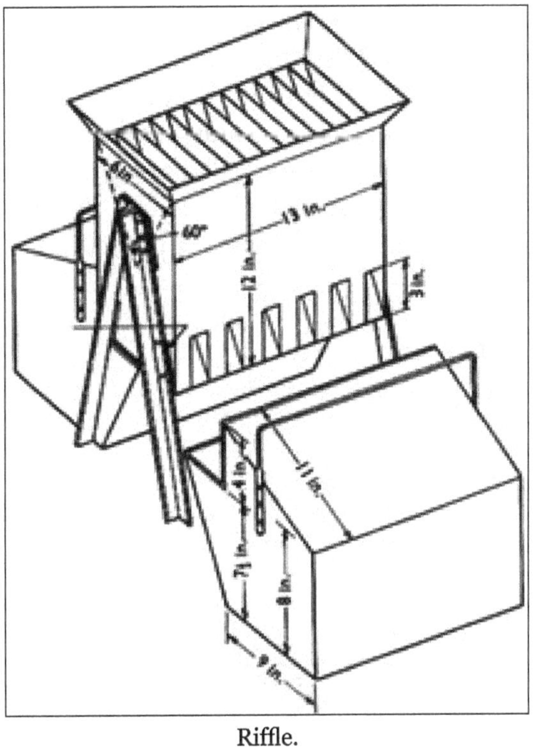

Riffle.

Examples:

1. A Consignment Comprising 200 Bags: The square root of 200=14.142, therefore n=14: makeup 14 group of 14 bags (i.e. total of 196 bags); Draw up a list from 1 to 14; cross out one number, for e.g. 7; sample the seventh bag from each group of 14 bags; the remaining group (i.e. 4) is smaller than 14 bags, so sample one bag from this group at random. A total of 15 bags have thereof been selected.

2. A Consignment Comprising 2000 Bags: The square root of 2000=44.721, therefore n=45: make up 44 groups of 45 bags (i.e. total of 1980 bags); draw up a list from 1 to 45; cross out one number, for example 20; sample the 20th bag from each group of 45 bags; the remaining group (i.e. 20) is smaller than 45 bags, so sample one bag from this group at random. A total of 45 bags ha therefore been selected.

3. Reduction of the Sample to Analytical Size: The two common methods of reducing bulk samples to a practical size are riffling and quartering. The sample is fed onto the top of the riffle and as it falls through the device the sample is divided equally into two bins. When the operation is complete,

the contents of one bin is discarded and the other passed through the riffle again. In this way the sample is progressively and randomly halved until its bulk is reduced to that required for laboratory work up. The riffle, due to its design is extremely difficult to clean and therefore it is practical to arrange for an individual riffle should be permanently kept for use with a specific material.

Sample Quartering

The total sample is placed on a clean impervious surface and formed into a cone by shoveling. Each shovel full is placed on the top of the last so that the material runs down the side of the pile and is thereby distributed around the sides of the pile as evenly as possible (if larger pieces of material roll away from the base of the cone they must be pushed back to the edge). From the first cone two successive cones are sequentially made in the same manner (this is to mix the sample well). The third cone is then flattened to a uniform thickness and then quartered along two diameters. One diagonal pair of quarters is rejected. These four stages are repeated until a sample of the required size is obtained. If the initial bulk sample is very course then the average particle diameter made need to be reduced by grinding between each quartering procedure.

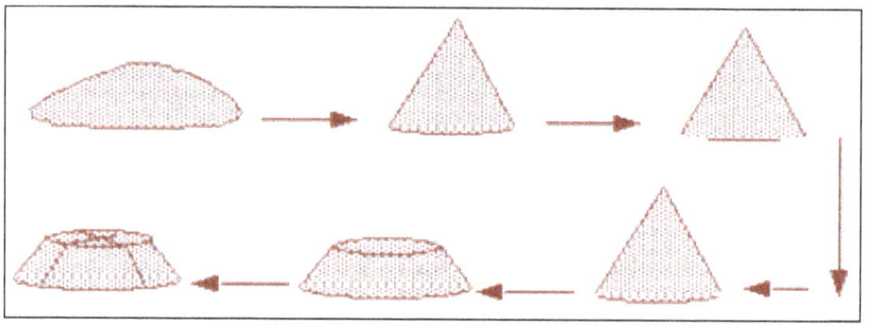

Instruments for Different Types of Product and Storage States

Sample Dividers

Ideal for Free Flowing Powders Suitable for use with powder chemicals, food stuff, feed and similar granular materials These hand held sample dividers will subdivide material samples into smaller portions by single or multiple passes. The important feature of Endecotts sample dividers is that each subdivision retains the characteristics of the original sample.

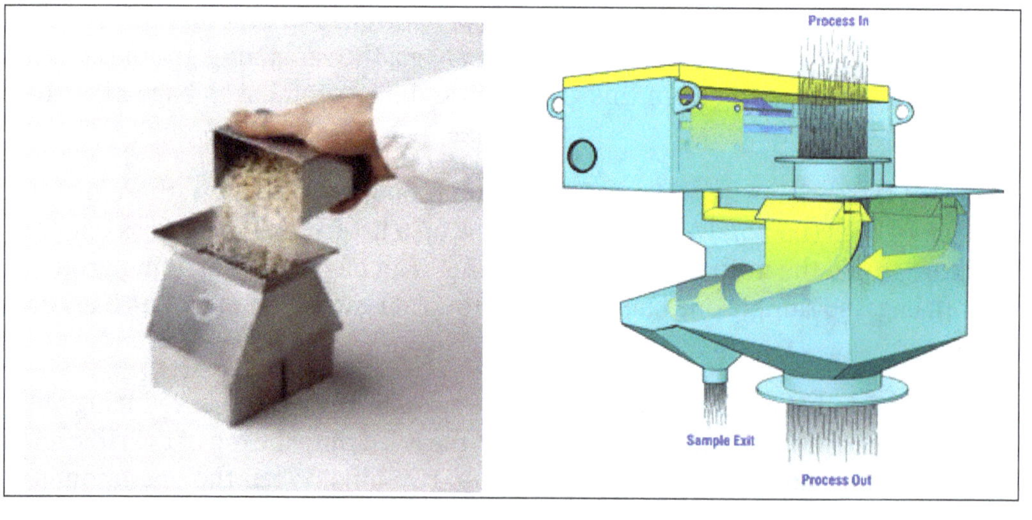

Sample Scoops

These Heavy Duty Sample Scoops are produced to the highest quality, are crevice free to minimise contamination and easy to clean.

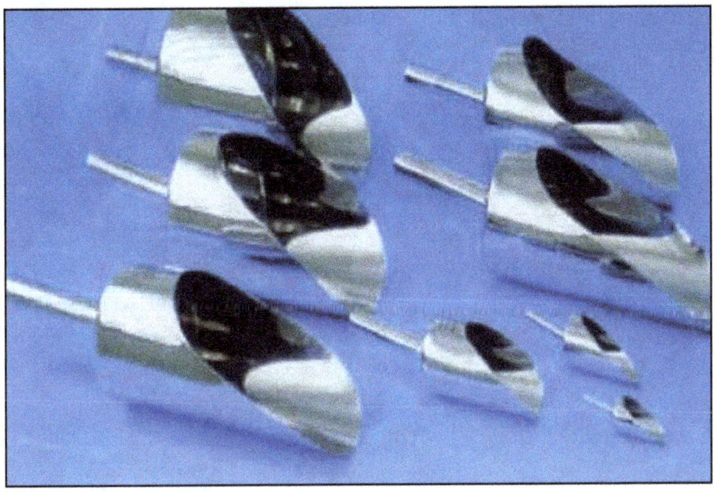

Sleeve Sampler

Ideal for Free Flowing Powders Suitable for use with powder chemicals, food stuff, feed and similar granular materials Ideal for sampling large volumes at great depths. At the required depth pull up the sampler slightly. This will force the sleeve down so that the product can fall into the sample chamber.

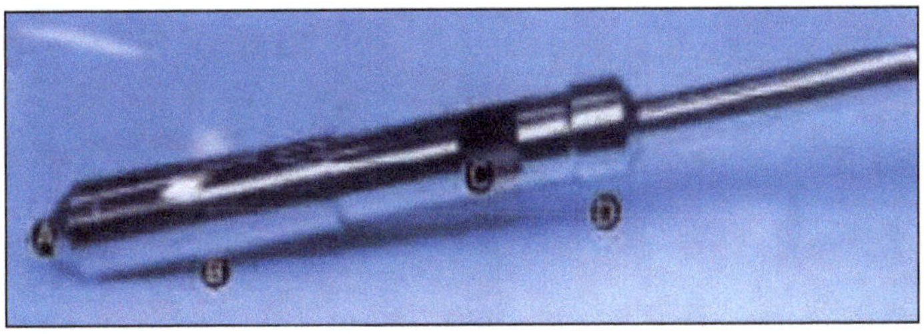

The Sampling Spear

Powder, Granules or Crystals

Samples of powders granules or crystals are usually taken with a spear. The spear is thrust at an angle into the material (the opening underneath) rotated two or three times and then carefully withdrawn with the opening uppermost. The contents are then discharged into the sample container.

References

- "Position of the American Dietetic Association: Fat replacers". Journal of the American Dietetic Association. 105 (2): 266–275. 2005. Doi:10.1016/j.jada.2004.12.011. PMID 15668687
- Factors-affecting-food-quality: blogspot.com, Retrieved 14 July, 2019

- "Micronutrient Fortification and Biofortification Challenge | Copenhagen Consensus Center". Www.copenhagenconsensus.com. Retrieved 2017-06-14
- Quality-management, services: campdenbri.co.uk, Retrieved 17 May, 2019
- Mcnulty, Helene; Pentieva, Kristina (2007). "Folate bioavailability". Proceedings of the Nutrition Society. 63 (4): 529–536. Doi:10.1079/PNS2004383. ISSN 0029-6651
- Saravacos, George D.; Maroulis, Zacharias B. (2011). Food Process Engineering Operations. CRC Press. Pp. 198-199. ISBN 1439877858

3

Methods of Food Safety

There are various methods and technologies that are used to ensure the safety of food. A few of its methods and technologies are pasteurization, pascalization, hurdle technology, food irradiation, etc. The topics elaborated in this chapter will help in gaining a better perspective about these methods of food safety.

Pasteurization

Pasteurization is a process whereby fluids such as wine and milk are heated for a predetermined time at a temperature that is below the boiling point of the liquid. The treatment kills any microorganisms that are in the fluid but does not alter the taste, appearance, or nutritive value of the fluid.

The process of pasteurization is named after the French chemist Louis Pasteur who is regarded as the founder of the study of modern microbiology. Among Pasteur's many accomplishments was the observation that the heating of fluids destroys harmful bacteria.

The basis of pasteurization is the application of heat. Many bacteria cannot survive exposure to the range of temperatures used in pasteurization. The energy of the heating process is disruptive to the membrane(s) that enclose the bacteria. As well, the bacterial enzymes that are vital for the maintenance of the growth and survival of the bacteria are denatured, or lose their functional shape, when exposed to heat. The disruption of bacteria is usually so complete that recovery of the cells following the end of the heat treatment is impossible.

The pasteurization process is a combination of temperature, time, and the consistency of the product. Thus, the actual conditions of pasteurization can vary depending on the product being treated. For example heating at 145 °F (63 °C) for not less than 30 minutes or at 162 °F (72 °C) for not less than 16 seconds pasteurizes milk. A product with greater consistency, such ice cream or egg nog, is pasteurized by heating at a temperature of at least 156 °F (69 °C) for not less than 30 minutes or at a temperature of at least 176 °F (80 °C) for not less than 25 seconds.

Particularly in commercial settings, such as a milk processing plant, there are two long-standing methods of pasteurization. These are known as the batch method and the continuous method. In the batch method the fluid is held in one container throughout the process. This method of pasteurization tends to be used for products such as ice cream. Milk tends to be pasteurized using the continuous method.

In the continuous method the milk passes by a stack of steel plates that are heated to the desired temperature. The flow rate is such that the milk is maintained at the desired temperature for the specified period of time. The pasteurized milk then flows to another tank.

Several other more recent variations on the process of pasteurization have been developed. The first of these variations is known as flash pasteurization. This process uses a higher temperature than conventional pasteurization, but the temperature is maintained for a shorter time. The product is then rapidly cooled to below 50 °F (10 °C), a temperature at which it can then be stored. The intent of flash pasteurization is to eliminate harmful microorganisms while maintaining the product as close as possible to its natural state. Juices are candidates for this process. In milk, lactic acid bacteria can survive. While these bacteria are not a health threat, their subsequent metabolic activity can cause the milk to sour.

Another variation on pasteurization is known as ultra-pasteurization. This is similar to flash pasteurization, except that a higher than normal pressure is applied. The higher pressure greatly increases the temperature that can be achieved, and so decreases the length of time that a product, typically milk, needs to be exposed to the heat. The advantage of ultra-pasteurization is the extended shelf live of the milk that results. The milk, which is essentially sterile, can be stored unopened at room temperature for several weeks without compromising the quality.

In recent years the term cold pasteurization has been used to describe the sterilization of solids, such as food, using radiation. The applicability of using the term pasteurization to describe a process that does not employ heat remains a subject of debate among microbiologists.

Pasteurization is effective only until the product is exposed to the air. Then, microorganisms from the air can be carried into the product and growth of microorganisms will occur. The chance of this contamination is lessened by storage of milk and milk products at the appropriate storage temperatures after they have been opened. For example, even ultra-pasteurized milk needs to stored in the refrigerator once it is in use.

Pascalization

Pascalization, bridgmanization, high pressure processing (HPP) or high hydrostatic pressure (HHP) processing is a method of preserving and sterilizing food, in which a product is processed under very high pressure, leading to the inactivation of certain microorganisms and enzymes in the food. HPP has a limited effect on covalent bonds within the food product, thus maintaining both the sensory and nutritional aspects of the product. The technique was named after Blaise Pascal, a French scientist of the 17th century whose work included detailing the effects of pressure on fluids. During pascalization, more than 50,000 pounds per square inch (340 MPa, 3.4 kbar) may be applied for around fifteen minutes, leading to the inactivation of yeast, mold, and bacteria. Pascalization is also known as bridgmanization, named for physicist Percy Williams Bridgman.

Uses

Spoilage microorganisms and some enzymes can be deactivated by HPP, which can extend the shelf life while preserving the sensory and nutritional characteristics of the product. Pathogenic microorganisms such as Listeria, E. coli, Salmonella, and Vibrio are also sensitive to pressures of 400-1000 MPa used during HPP. Thus, HPP can pasteurize food products with decreased processing time, reduced energy usage, and less waste. The treatment occurs at low temperatures and does not include the use of food additives. From 1990, some juices, jellies, and jams have been preserved using pascalization in Japan. The technique is now used there to preserve fish and meats, salad dressing, rice cakes, and yogurts. HPP is now being used to preserve fruit and vegetable smoothies and other products such as meat for sale in the UK. An early use of pascalization in the United States was to treat guacamole. It did not change the guacamole's taste, texture, or color, but the shelf life of the product increased to thirty days, from three days without the treatment. However, some treated foods still require cold storage because pascalization does not stop all enzyme activity caused by proteins, some of which affects shelf life. In recent years, HPP has also been used in the processing of raw pet food. Most commercial frozen and freeze-dried raw diets now go through post-packaging HPP treatment to destroy potential bacteria and viruses contaminants, with salmonella being one of the biggest concerns.

Process

In pascalization, food products are sealed and placed into a steel compartment containing a liquid, often water, and pumps are used to create pressure. The pumps may apply pressure constantly or intermittently. The application of high hydrostatic pressures (HHP) on a food product will kill many microorganisms, but the spores are not destroyed. Pascalization works especially well on acidic foods, such as yogurts and fruits, because pressure-tolerant spores are not able to live in environments with low pH levels. The treatment works equally well for both solid and liquid products.

Bacterial spores survive pressure treatment at ambient or chilled conditions. Researchers reported that pressure in combination with heat is effective in the inactivation of bacterial spores. The process is called pressure-assisted thermal sterilization. In 2009 and 2015, Food and Drug Administration (FDA) issued letters of no objection for two industrial petition for pressure-assisted thermal processing. At this time, there are no commercial low-acid products treated by PATP are available in the market.

During pascalization, the food's hydrogen bonds are selectively disrupted. Because pascalization is not heat-based, covalent bonds are not affected, causing no change in the food's taste. This means that HPP does not destroy vitamins, maintaining the nutritional value of the food. High hydrostatic pressure can affect muscle tissues by increasing the rate of lipid oxidation, which in turn leads to poor flavor and decreased health benefits. Additionally, there are some compounds present in foods that are subject to change during the treatment process. For example, carbohydrates are gelatinized by an increase in pressure instead of increasing the temperature during the treatment process.

Because hydrostatic pressure is able to act quickly and evenly on food, neither the size of a product's container nor its thickness play a role in the effectiveness of pascalization. There are several side effects of the process, including a slight increase in a product's sweetness, but pascalization does not greatly affect the nutritional value, taste, texture, and appearance. As a result, high pressure treatment of foods is regarded as a "natural" preservation method, as it does not use chemical preservatives.

Anurag Sharma, a geochemist, James Scott, a microbiologist, and others at the Carnegie Institution of Washington directly observed microbial activity at pressures in excess of 1 gigapascal. The experiments were performed up to 1.6 GPa (232,000 psi) of pressure, which is more than 16,000 times normal air pressure, or about 14 times the pressure in the deepest ocean trench.

The experiment began by depositing an *Escherichia coli* and *Shewanella oneidensis* film in a Diamond Anvil Cell (DAC). The pressure was then raised to 1.6 GPa. When raised to this pressure and kept there for 30 hours, at least 1% of the bacteria survived. The experimenters then monitored formate metabolism using in-situ Raman spectroscopy and showed that formate metabolism continued in the bacterial sample.

Moreover, 1.6 GPa is such great pressure that during the experiment the DAC turned the solution into ice-IV, a room-temperature ice. When the bacteria broke down the formate in the ice, liquid pockets would form because of the chemical reaction.

There was some skepticism of this experiment. According to Art Yayanos, an oceanographer at the Scripps Institute of Oceanography, an organism should only be considered living if it can reproduce. Another issue with the DAC experiment is that when high pressures occur, there are usually high temperatures present as well, but in this experiment there were not. This experiment was performed at room-temperature. However, the intentional lack of high temperature in the experiments isolated the actual effects of pressure on life and results clearly indicated life to be largely pressure insensitive.

Newer results from independent research groups have confirmed Sharma et al. This is a significant step that reiterates the need for a new approach to the old problem of studying environmental extremes through experiments. There is practically no debate whether microbial life can survive pressures up to 600 MPa, which has been shown over the last decade or so to be valid through a number of scattered publications.

Consumer Acceptance of Pascalization

In the consumer studies of Hightech Europe consumers mentioned more positive than negative associations descriptions for this technology showing that these products are well accepted.

Hurdle Technology

Hurdle technology is a method of ensuring that pathogens in food products can be eliminated or controlled. This means the food products will be safe for consumption, and their shelf life will be extended. Hurdle technology usually works by combining more than one approach. These approaches can be thought of as "hurdles" the pathogen has to overcome if it is to remain active in the food. The right combination of hurdles can ensure all pathogens are eliminated or rendered harmless in the final product.

Hurdle technology has been defined by Leistner as an intelligent combination of hurdles which secures the microbial safety and stability as well as the organoleptic and nutritional quality and the economic viability of food products. The organoleptic quality of the food refers to its sensory properties, that is its look, taste, smell and texture.

Examples of hurdles in a food system are high temperature during processing, low temperature during storage, increasing the acidity, lowering the water activity or redox potential, or the presence of preservatives. According to the type of pathogens and how risky they are, the intensity of the hurdles can be adjusted individually to meet consumer preferences in an economical way, without compromising the safety of the product.

Hurdles

Each hurdle aims to eliminate, inactivate or at least inhibit unwanted microorganisms. Common salt or organic acids can be used as hurdles to control microbials in food. Many natural antimicrobials such as nisin, natamycin and other bacteriocins, and essential oils derived from rosemary or thyme, also work well.

Principal hurdles used for food preservation.		
Parameter	Symbol	Application
High temperature	F	Heating
Low temperature	T	Chilling, freezing
Reduced water activity	a_w	Drying, curing, conserving
Increased acidity	pH	Acid addition or formation
Reduced redox potential	E_h	Removal of oxygen or addition of ascorbate
Biopreservatives		Competitive flora such as microbial fermentation
Other preservatives		Sorbates, sulfites, nitrites

"Traditionally, fermented seafood products common in Japan, provide a typical example of hurdle technology. Fermentation of sushi employs hurdles that favour growth of desirable bacteria but inhibit the growth of pathogens. The important hurdles in the early stages of fermentation are salt and vinegar. Raw fish is cured in salt (20–30%, w/w) for one month before being desalted and pickled in vinegar. The main target of these hurdles is C. botulinum. Growth of lactic acid bacteria during fermentation results in acid production from metabolism of added sugars and rice. The result is a pH hurdle important in controlling growth of C. botulinum."

Types of hurdles used for food preservation (from Ohlsson and Bengtsson, 2002)	
Type of hurdle	Examples
Physical	Aseptic packaging, electromagnetic energy (microwave, radio frequency, pulsed magnetic fields, high electric fields), high temperatures (blanching, pasteurization, sterilization, evaporation, extrusion, baking, frying), ionizing radiation, low temperature (chilling, freezing), modified atmospheres, packaging films (including active packaging, edible coatings), photodynamic inactivation, ultra-high pressures, ultrasonication, ultraviolet radiation.
Physicochemical	Carbon dioxide, ethanol, lactic acid, lactoperoxidase, low pH, low redox potential, low water activity, Maillard reaction products, organic acids, oxygen, ozone, phenols, phosphates, salt, smoking, sodium nitrite/nitrate, sodium or potassium sulphite, spices and herbs, surface treatment agents.
Microbial	Antibiotics, bacteriocins, competitive flora, protective cultures.

Synergistic Effects

There can be significant synergistic effects between hurdles. For example, Gram-positive bacteria include some of the more important spoilage bacteria, such as Clostridium, Bacillus and Listeria. A synergistic enhancement occurs if nisin is used against these bacteria in combination with antioxidants, organic acids or other antimicrobials. Combining antimicrobial hurdles in an intelligent way means other hurdles can be reduced, yet the resulting food can have superior sensory qualities.

Food Irradiation

Food irradiation is the process of exposing food and food packaging to ionizing radiation. Ionizing radiation, such as from gamma rays, x-rays, or electron beams, is energy that can be transmitted without direct contact to the source of the energy (radiation) capable of freeing electrons from their atomic bonds (ionization) in the targeted food. The radiation can be emitted by a radioactive substance or generated electrically. This treatment is used to improve food safety by extending product shelf-life (preservation), reducing the risk of foodborne illness, delaying or eliminating sprouting or ripening, by sterilization of foods, and as a means of controlling insects and invasive pests. Food irradiation primarily extends the shelf-life of irradiated foods by effectively destroying organisms responsible for spoilage and foodborne illness and inhibiting sprouting.

Cobalt-60 irradiation facility is used to test irradiation as a tool to ensure food safety.

Although consumer perception of foods treated with irradiation is more negative than those processed by other means, because people imagine that the food is radioactive or mutated, these thoughts don't agree with the understood mechanism by which irradiation works. The food itself is already not alive, so irradiation will not affect it meaningfully. Irradiation will kill the living bacteria, however. Additionally, all independent research, the U.S. Food and Drug Administration (FDA), the World Health Organization (WHO), the Centers for Disease Control and Prevention (CDC), and U.S. Department of Agriculture (USDA) have performed studies that confirm irradiation to be safe. In order for a food to be irradiated in the US, the FDA will still require that the specific food be thoroughly tested for irradiation safety.

The international Radura logo, used to show a food has been treated with ionizing radiation.

Food irradiation is permitted by over 60 countries, with about 500,000 metric tons of food annually processed worldwide. The regulations that dictate how food is to be irradiated, as well as the food allowed to be irradiated, vary greatly from country to country. In Austria, Germany, and many other countries of the European Union only dried herbs, spices, and seasonings can be processed with irradiation and only at a specific dose, while in Brazil all foods are allowed at any dose.

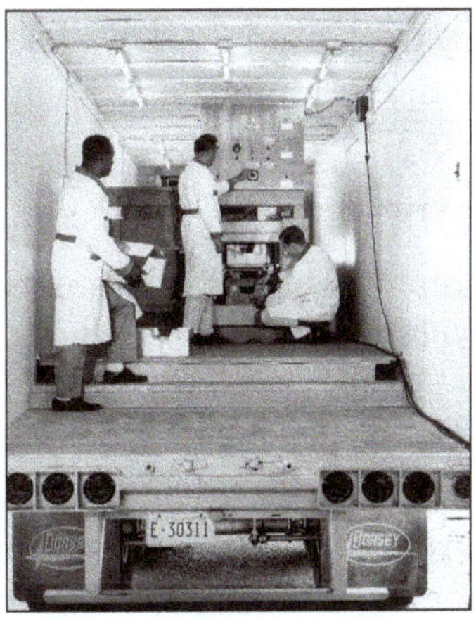

A portable, trailer-mounted food irradiation machine, circa 1968.

Uses

Irradiation is used to reduce or eliminate the risk of food-borne illnesses, prevent or slow down spoilage, arrest maturation or sprouting and as a treatment against pests. Depending on the dose, some or all of the pathogenic organisms, microorganisms, bacteria, and viruses present are destroyed, slowed down, or rendered incapable of reproduction. Irradiation cannot return spoiled or over-ripe food to a fresh state. If this food was processed by irradiation, further spoilage would cease and ripening would slow down, yet the irradiation would not destroy the toxins or repair the texture, color, or taste of the food. When targeting bacteria, most foods are irradiated to significantly reduce the number of active microbes, not to sterilize all microbes in the product. In this respect it is similar to pasteurization.

Irradiation is used to create safe foods for people at high risk of infection, or for conditions where food must be stored for long periods of time and proper storage conditions are not available. Foods that can tolerate irradiation at sufficient doses are treated to ensure that the product is completely sterilized. This is most commonly done with rations for astronauts, and special diets for hospital patients.

Irradiation is used to create shelf-stable products. Since irradiation reduces the populations of spoilage microorganisms, and because pre-packed food can be irradiated, the packaging prevents recontamination of the final product.

Irradiation is used to reduce post-harvest losses. It reduces populations of spoilage micro-organisms in the food and can slow down the speed at which enzymes change the food, and therefore slows spoilage and ripening, and inhibits sprouting (e.g., of potato, onion, and garlic).

Food is also irradiated to prevent the spread of invasive pest species through trade in fresh vegetables and fruits, either within countries, or trade across international boundaries. Pests such as insects could be transported to new habitats through trade in fresh produce which could significantly affect agricultural production and the environment were they to establish themselves. This "phytosanitary irradiation" aims to render any hitch-hiking pest incapable of breeding. The pests are sterilized when the food is treated by low doses of irradiation. In general, the higher doses required to destroy pests such as insects, mealybugs, mites, moths, and butterflies either affect the look or taste, or cannot be tolerated by fresh produce. Low dosage treatments (less than 1000 gray) enables trade across quarantine boundaries and may also help reduce spoilage.

Impact

Irradiation reduces the risk of infection and spoilage, does not make food radioactive, and the food is shown to be safe, but it does cause chemical reactions that alter the food and therefore alters the chemical makeup, nutritional content, and the sensory qualities of the food. Some of the potential secondary impacts of irradiation are hypothetical, while others are demonstrated. These effects include cumulative impacts to pathogens, people, and the environment due to the reduction of food quality, the transportation and storage of radioactive goods, and destruction of pathogens, changes in the way we relate to food and how irradiation changes the food production and shipping industries.

Immediate Effects

The radiation source supplies energetic particles or waves. As these waves/particles pass through a target material they collide with other particles. Around the sites of these collisions chemical bonds are broken, creating short lived radicals (e.g. the hydroxyl radical, the hydrogen atom and solvated electrons). These radicals cause further chemical changes by bonding with and or stripping particles from nearby molecules. When collisions damage DNA or RNA, effective reproduction becomes unlikely, also when collisions occur in cells, cell division is often suppressed.

Irradiation (within the accepted energy limits, as 10 MeV for electrons, 5 MeV for X-rays [US 7.5 MeV] and gamma rays from Cobalt-60) can not make food radioactive, but it does produce radiolytic products, and free radicals in the food. A few of these products are unique, but not considered dangerous.

Irradiation can also alter the nutritional content and flavor of foods, much like cooking. The scale of these chemical changes is not unique. Cooking, smoking, salting, and other less novel techniques, cause the food to be altered so drastically that its original nature is almost unrecognizable, and must be called by a different name. Storage of food also causes dramatic chemical changes, ones that eventually lead to deterioration and spoilage.

Misconceptions

A major concern is that irradiation might cause chemical changes that are harmful to the consumer. Several national expert groups and two international expert groups evaluated the available data and concluded that any food at any dose is wholesome and safe to consume as long as it remains palatable and maintains its technical properties (e.g. feel, texture, or color).

Irradiated food does not become radioactive, just as an object exposed to light does not start producing light. Radioactivity is the ability of a substance to emit high energy particles. When particles hit the target materials they may free other highly energetic particles. This ends shortly after the end of the exposure, much like objects stop reflecting light when the source is turned off and warm objects emit heat until they cool down but do not continue to produce their own heat. To modify a material so that it keeps emitting radiation (induce radiation) the atomic cores (nucleus) of the atoms in the target material must be modified.

It is impossible for food irradiators to induce radiation in a product. Irradiators emit electrons or photons and the radiation is intrinsically radiated at precisely known strengths (wavelengths for photons, and speeds for electrons). These radiated particles at these strengths can never be strong enough to modify the nucleus of the targeted atom in the food, regardless of how many particles hit the target material, and radioactivity can not be induced without modifying the nucleus.

Chemical Changes

Compounds known as free radicals form when food is irradiated. Most of these are oxidizers (i.e., accept electrons) and some react very strongly. According to the free-radical theory of aging excessive amounts of these free radicals can lead to cell injury and cell death, which may contribute to many diseases. However, this generally relates to the free radicals generated in the body, not the free radicals consumed by the individual, as much of these are destroyed in the digestive process.

Most of the substances found in irradiated food are also found in food that has been subjected to other food processing treatments, and are therefore not unique. One family of chemicals (2ACB's) are uniquely formed by irradiation (unique radiolytic products), and this product is nontoxic. When fatty acids are irradiated, a family of compounds called 2-alkylcyclobutanones (2-ACBs) are produced. These are thought to be unique radiolytic products. When irradiating food, all other chemicals occur in a lower or comparable frequency to other food processing techniques. Furthermore, the quantities in which they occur in irradiated food are lower or similar to the quantities formed in heat treatments.

The radiation doses to cause toxic changes are much higher than the doses used during irradiation, and taking into account the presence of 2-ACBs along with what is known of free radicals, these results lead to the conclusion that there is no significant risk from radiolytic products.

Food Quality

Ionizing radiation can change food quality but in general very high levels of radiation treatment (many thousands of gray) are necessary to adversely change nutritional content, as well as the sensory qualities (taste, appearance, and texture). Irradiation to the doses used commercially to treat food have very little negative impact on the sensory qualities and nutrient content in foods. When irradiation is used to maintain food quality for a longer period of time (improve the shelf stability of some sensory qualities and nutrients) the improvement means that more consumers have access to the original taste, texture, appearance, and nutrients. The changes in quality and nutrition depend on the degree of treatment and may vary greatly from food to food.

There has been low level gamma irradiation that has been attempted on arugula, spinach, cauliflower, ash gourd, bamboo shoots, coriander, parsley, and watercress. There has been limited information, however, regarding the physical, chemical and bioactive properties and the shelf life on these minimally processed vegetables.

There is some degradation of vitamins caused by irradiation, but is similar to or even less than the loss caused by other processes that achieve the same result. Other processes like chilling, freezing, drying, and heating also result in some vitamin loss.

The changes in the flavor of fatty foods like meats, nuts and oils are sometimes noticeable, while the changes in lean products like fruits and vegetables are less so. Some studies by the irradiation industry show that for some properly treated fruits and vegetables irradiation is seen by consumers to improve the sensory qualities of the product compared to untreated fruits and vegetables.

Quality Impact on Minimally Processed Vegetables

Watercress (Nasturtium Officinale) is a rapidly growing aquatic or semi aquatic perennial plant. Because chemical agents do not provide efficient microbial reductions, watercress has been tested with gamma irradiation treatment in order to improve both safety and the shelf life of the product. It is traditionally used on horticultural products to prevent sprouting and post-packaging contamination, delay post-harvest ripening, maturation and senescence.

In a Food Chemistry food journal, scientists studied the suitability of gamma irradiation of 1, 2, and 5 kGy for preserving quality parameters of the fresh cut watercress at around 4 degrees Celsius for 7 days. They determined that a 2 kGy dose of irradiation was the dose that contained most similar qualities to non-stored control samples, which is one of the goals of irradiation. 2 kGy preserved high levels of reducing sugars and favoured polyunsaturated fatty acids (PUFA); while samples of the 5 kGy dose revealed high contents of sucrose and monounsaturated fat (MUFA). Both cases the watercress samples obtained healthier fatty acids profiles. However, a 5kGy dose better preserved the antioxidant activity and total flavonoids.

Long-term Impacts

If the majority of food was irradiated at high-enough levels to significantly decrease its nutritional content, there would be an increased risk of developing nutritionally-based illnesses if additional steps, such as changes in eating habits, were not taken to mitigate this. Furthermore, for at least three studies on cats, the consumption of irradiated food was associated with a loss of tissue in the myelin sheath, leading to reversible paralysis. Researchers suspect that reduced levels of vitamin A and high levels of free radicals may be the cause. This effect is thought to be specific to cats and has not been reproduced in any other animal. To produce these effects, the cats were fed solely on food that was irradiated at a dose at least five times higher than the maximum allowable dose.

permitted in the US for commercial food items by the FDA and other regulators around the world. Though these doses are approved for non commercial applications, such as sterilizing frozen meat for NASA astronauts (doses of 44 kGy) and food for hospital patients.

Applications of food irradiation		
	Application	Dose (kGy)
Low dose (up to 1 kGy)	Inhibit sprouting (potatoes, onions, yams, garlic)	0.06 - 0.2
	Delay in ripening (strawberries, potatoes)	0.5 - 1.0
	Prevent insect infestation (grains, cereals, coffee beans, spices, dried nuts, dried fruits, dried fish, mangoes, papayas)	0.15 - 1.0
	Parasite control and inactivation (tape worm, trichina)	0.3 - 1.0
Medium dose (1 kGy to 10 kGy)	Extend shelf-life (raw and fresh fish, seafood, fresh produce, refrigerated and frozen meat products)	1.0 - 7.0
	Reduce risk of pathogenic and spoilage microbes (meat, seafood, spices, and poultry)	1.0 - 7.0
	Increased juice yield, reduction in cooking time of dried vegetables	3.0 - 7.0
High dose (above 10 kGy)	Enzymes (dehydrated)	10.0
	Sterilization of spices, dry vegetable seasonings	30.0 max
	Sterilization of packaging material	10.0 - 25.0
	Sterilization of foods (NASA and hospitals)	44.0

Processes

Gamma Irradiation

Gamma irradiation is produced from the radioisotopes cobalt-60 and caesium-137, which are derived by neutron bombardment of cobalt-59 and as a nuclear source by-product, respectively. Cobalt-60 is the most common source of gamma rays for food irradiation in commercial scale facilities as it is water insoluble and hence has little risk of environmental contamination by leakage into the water systems. As for transportation of the radiation source, cobalt-60 is transported in special trucks that prevent release of radiation and meet standards mentioned in the Regulations for Safe Transport of Radioactive Materials of the International Atomic Energy Act. The special trucks must meet high safety standards and pass extensive tests to be approved to ship radiation sources. Conversely, caesium-137, is water soluble and poses a risk of environmental contamination. Insufficient quantities are available for large scale commercial use. An incident where water-soluble caesium-137 leaked into the source storage pool requiring NRC intervention has led to near elimination of this radioisotope.

Cobalt 60 stored in Gamma Irradiation machine.

Gamma irradiation is widely used due to its high penetration depth and dose uniformity, allowing for large-scale applications with high through puts. Additionally, gamma irradiation is significantly less expensive than using an X-ray source In most designs, the radioisotope, contained in stainless steel pencils, is stored in a water-filled storage pool which absorbs the radiation energy when not in use. For treatment, the source is lifted out of the storage tank, and product contained in totes is passed around the pencils to achieve required processing.

Electron Beam

Treatment of electron beams is created as a result of high energy electrons in an accelerator that generates electrons accelerated to 99% the speed of light. This system uses electrical energy and can be powered on and off. The high power correlates with a higher throughput and lower unit cost, but electron beams have low dose uniformity and a penetration depth of centimeters. Therefore, electron beam treatment works for products that have low thickness.

Irradiated Guava: Spring Valley Fruits.

X-ray

X-rays are produced by bombardment of dense target material with high energy accelerated electrons (this process is known as bremsstrahlung-conversion), giving rise to a continuous energy spectrum. Heavy metals, such as tantalum and tungsten, are used because of their high atomic numbers and high melting temperatures. Tantalum is usually preferred versus tungsten for industrial, large-area, high-power targets because it is more workable than tungsten and has a higher threshold energy for induced reactions. Like electron beams, x-rays do not require the use of radioactive materials and can be turned off when not in use. X-rays have high penetration depths and high dose uniformity but they are a very expensive source of irradiation as only 8% of the incident energy is converted into X-rays.

Cost

The cost of food irradiation is influenced by dose requirements, the food's tolerance of radiation, handling conditions, i.e., packaging and stacking requirements, construction costs, financing arrangements, and other variables particular to the situation. Irradiation is a capital-intensive technology requiring a substantial initial investment, ranging from $1 million to $5 million. In the case of large research or contract irradiation facilities, major capital costs include a radiation source, hardware (irradiator, totes and conveyors, control systems, and other auxiliary equipment), land (1 to 1.5 acres), radiation shield, and warehouse. Operating costs include salaries (for fixed and variable labor), utilities, maintenance, taxes/insurance, cobalt-60 replenishment, general utilities, and miscellaneous operating costs. Perishable food items, like fruits, vegetables and meats would still require to be handled in the cold chain, so all other supply chain costs remain the same.

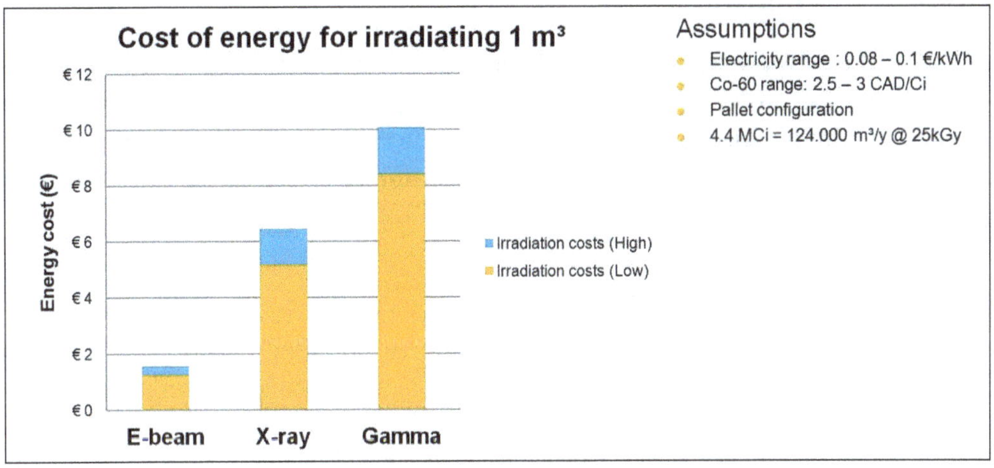

Efficiency illustration of the different radiation technologies (electron beam, X-ray, gamma rays).

Public Perception

Negative connotations associated with the word "radiation" are thought to be responsible for low consumer acceptance. Several national expert groups and two international expert groups evaluated the available data and concluded that any food at any dose is wholesome and safe to consume.

Irradiation has been approved by many countries. For example, in the U.S. the FDA has approved food irradiation for over fifty years. However, in the past decade the major growth area is for fruits and vegetables that are irradiated to prevent the spread of pests. In the early 2000s in the US, irradiated meat was common at some grocery stores, but because of lack of consumer demand, it is no longer common. Because consumer demand for irradiated food is low, reducing the spoilage between manufacturer and consumer purchase and reducing the risk of food borne illness is currently not sufficient incentive for most manufacturers to supplement their process with irradiation. Nevertheless, food irradiation does take place commercially and volumes are in general increasing at a slow rate, even in the European Union where all member countries allow the irradiation of dried herbs spices and vegetable seasonings but only a few allow other foods to be sold as irradiated.

Although there are some consumers who choose not to purchase irradiated food, a sufficient market has existed for retailers to have continuously stocked irradiated products for years. When labeled irradiated food is offered for retail sale, these consumers buy it and re-purchase it, indicating that it is possible to successfully market irradiated foods, therefore retailers not stocking irradiated foods might be a major bottleneck to the wider adoption of irradiated foods. It is however, widely believed that consumer perception of foods treated with irradiation is more negative than those processed by other means and some industry studies indicate the number of consumers concerned about the safety of irradiated food decreased between 1985 and 1995 to levels comparable to those of people concerned about food additives and preservatives. Even though is it is untrue "People think the product is radioactive," said Harlan Clemmons, president of Sadex, a food irradiation company based in Sioux City, Iowa. Because of these concerns and the increased cost of irradiated foods, there is not a widespread public demand for the irradiation of foods for human consumption. Irradiated food does not become radioactive.

Standards and Regulations

The Codex Alimentarius represents the global standard for irradiation of food, in particular under the WTO-agreement. Regardless of treatment source, all processing facilities must adhere to safety standards set by the International Atomic Energy Agency (IAEA), Codex Code of Practice for the Radiation Processing of Food, Nuclear Regulatory Commission (NRC), and the International Organization for Standardization (ISO). More specifically, ISO 14470 and ISO 9001 provide in-depth information regarding safety in irradiation facilities.

All commercial irradiation facilities contain safety systems are designed to prevent exposure of personnel to radiation. The radiation source is constantly shielded by water, concrete, or metal. Irradiation facilities are designed with overlapping layers of protection, interlocks, and safeguards to prevent accidental radiation exposure. Additionally, "melt-downs" do not occur in facilities because the radiation source gives off radiation and decay heat; however, the heat is not sufficient to melt any material.

Labeling

The Radura symbol, as required by U.S. Food and Drug Administration regulations to show a food has been treated with ionizing radiation.

The provisions of the Codex Alimentarius are that any "first generation" product must be labeled "irradiated" as any product derived directly from an irradiated raw material; for ingredients the provision is that even the last molecule of an irradiated ingredient must be listed with the ingredients even in cases where the unirradiated ingredient does not appear on the label. The RADURA-logo is optional; several countries use a graphical version that differs from the Codex-version. The suggested rules for labeling is published at CODEX-STAN – 1 and includes the usage of the Radura symbol for all products that contain irradiated foods. The Radura symbol is not a designator of quality. The amount of pathogens remaining is based upon dose and the original content and the dose applied can vary on a product by product basis.

The European Union follows the Codex's provision to label irradiated ingredients down to the last molecule of irradiated food. The European Community does not provide for the use of the Radura logo and relies exclusively on labeling by the appropriate phrases in the respective languages of the Member States. The European Union enforces its irradiation labeling laws by requiring its member countries to perform tests on a cross section of food items in the market-place and to report to the European Commission. The results are published annually in the OJ of the European Communities.

The US defines irradiated foods as foods in which the irradiation causes a material change in the food, or a material change in the consequences that may result from the use of the food. Therefore, food that is processed as an ingredient by a restaurant or food processor is exempt from the labeling requirement in the US. All irradiated foods must include a prominent Radura symbol followed in addition to the statement "treated with irradiation" or "treated by irradiation". Bulk foods must be individually labeled with the symbol and statement or, alternatively, the Radura and statement should be located next to the sale container.

Packaging

Under section 409 of the Federal Food, Drug, and Cosmetic Act, irradiation of prepackaged foods requires premarket approval for not only the irradiation source for a specific food but also for the food packaging material. Approved packaging materials include various plastic films, yet does not cover a variety of polymers and adhesive based materials that have been found to meet specific standards. The lack of packaging material approval limits manufacturers production and expansion of irradiated prepackaged foods.

Approved materials by FDA for Irradiation according to 21 CFR 179.45:

Material	Paper (kraft)	Paper (glassine)	Paper-board	Cellophane (coated)	Polyolefin film	Polyesty-rene film	Ny-lon-6	Vegetable Parchment	Nylon 11
Irradiation (kGy)	.05	10	10	10	10	10	10	60	60

Food Safety

In 2003, the Codex Alimentarius removed any upper dose limit for food irradiation as well as clearances for specific foods, declaring that all are safe to irradiate. Countries such as Pakistan and Brazil have adopted the Codex without any reservation or restriction.

Standards that describe calibration and operation for radiation dosimetry, as well as procedures to relate the measured dose to the effects achieved and to report and document such results, are maintained by the American Society for Testing and Materials (ASTM international) and are also available as ISO/ASTM standards.

All of the rules involved in processing food are applied to all foods before they are irradiated.

United States

The U.S. Food and Drug Administration (FDA) is the agency responsible for regulation of radiation sources in the United States. Irradiation, as defined by the FDA is a "food additive" as opposed to a food process and therefore falls under the food additive regulations. Each food approved for irradiation has specific guidelines in terms of minimum and maximum dosage as determined safe by the FDA. Packaging materials containing the food processed by irradiation must also undergo approval. The United States Department of Agriculture (USDA) amends these rules for use with meat, poultry, and fresh fruit.

The United States Department of Agriculture (USDA) has approved the use of low-level irradiation as an alternative treatment to pesticides for fruits and vegetables that are considered hosts to a number of insect pests, including fruit flies and seed weevils. Under bilateral agreements that allows less-developed countries to earn income through food exports agreements are made to allow them to irradiate fruits and vegetables at low doses to kill insects, so that the food can avoid quarantine.

European Union

European law dictates that all member countries must allow the sale of irradiated dried aromatic herbs, spices and vegetable seasonings. However, these Directives allow Member States to maintain previous clearances food categories the EC's Scientific Committee on Food (SCF) had previously approved (the approval body is now the European Food Safety Authority). Presently, Belgium, Czech Republic, France, Italy, Netherlands, Poland, and the United Kingdom allow the sale of many different types of irradiated foods. Before individual items in an approved class can be added to the approved list, studies into the toxicology of each of such food and for each of the proposed dose ranges are requested. It also states that irradiation shall not be used "as a substitute for hygiene or health practices or good manufacturing or agricultural practice". These Directives only control food irradiation for food retail and their conditions and controls are not applicable to the irradiation of food for patients requiring sterile diets.

Because of the Single Market of the EC any food, even if irradiated, must be allowed to be marketed in any other Member State even if a general ban of food irradiation prevails, under the condition that the food has been irradiated legally in the state of origin. Furthermore, imports into the EC are possible from third countries if the irradiation facility had been inspected and approved by the EC and the treatment is legal within the EC or some Member state.

Australia

Australia banned irradiated cat food after a national scare where cats suffered from paralyzation after eating a specific brand of highly irradiated catfood for an extended period of time. The suspected culprit was malnutrition from consuming food depleted of Vitamin A by the irradiation process. The incident was linked only to a single batch of one brand's product and no illness was linked to any of that brand's other irradiated batches of the same product or to any other brand of irradiated cat food. This, along with incomplete evidence indicating that the cat food was not sufficiently depleted of Vitamin A makes irradiation a less likely cause. Further research has been able to experimentally induce the paralyzation of cats by via Vitamin A deficiency by feeding highly irradiated food.

Nuclear Safety and Security

Interlocks and safeguards are mandated to minimize this risk. There have been radiation-related accidents, deaths, and injury at such facilities, many of them caused by operators overriding the safety related interlocks. In a radiation processing facility, radiation specific concerns are supervised by special authorities, while "Ordinary" occupational safety regulations are handled much like other businesses.

The safety of irradiation facilities is regulated by the United Nations International Atomic Energy Agency and monitored by the different national Nuclear Regulatory Commissions. The regulators enforce a safety culture that mandates that all incidents that occur are documented and thoroughly analyzed to determine the cause and improvement potential. Such incidents are studied by personnel at multiple facilities, and improvements are mandated to retrofit existing facilities and future design.

In the US the Nuclear Regulatory Commission (NRC) regulates the safety of the processing facility, and the United States Department of Transportation (DOT) regulates the safe transport of the radioactive sources.

Irradiated Food Supply

As of 2010, the quantities of foods irradiated in Asia, the EU and the US were 285,200, 9,300, and 103,000 tons. Authorities in some countries use tests that can detect the irradiation of food items to enforce labeling standards and to bolster consumer confidence. The European Union monitors the market to determine the quantity of irradiated foods, if irradiated foods are labeled as irradiated, and if the irradiation is performed at approved facilities.

Irradiation of fruits and vegetables to prevent the spread of pest and diseases across borders has been increasing globally. In 2010, 18,446 tonnes of fruits and vegetables were irradiated in six countries for export quarantine control. 97% of this was exported to the United States.

In total, 103,000 tonnes of food products were irradiated on mainland United States in 2010. The three types of foods irradiated the most were spices (77.7%), fruits and vegetables (14.6%) and meat and poultry (7.77%). 17,953 tonnes of irradiated fruits and vegetables were exported to the mainland United States. Mexico, the United States' state of Hawaii, Thailand, Vietnam and India export irradiated produce to the mainland U.S. Mexico, followed by the United States' state of Hawaii, is the largest exporter of irradiated produce to the mainland U.S.

In total, 6,876 tonnes of food products were irradiated in European Union countries in 2013,mainly in four member state countries: Belgium (49.4%), the Netherlands (24.4%), Spain (12.7%) and France (10.0%). The two types of foods irradiated the most were frog legs (46%), and dried herbs and spices (25%). There has been a decrease of 14% in the total quantity of products irradiated in the EU compared to the previous year 2012 (7,972 tonnes).

The U.S. Food and Drug Administration and the U.S. Department of Agriculture have approved irradiation of the following foods and purposes:

- Packaged refrigerated or frozen red meat: To control pathogens (E. Coli O157:H7 and Salmonella) and to extend shelf life.
- Packaged poultry: Control pathogens (Salmonella and Camplylobacter).
- Fresh fruits, vegetables, and grains: To control insects and inhibit growth, ripening and sprouting.
- Pork: To control trichinosis.

- Herbs, spices and vegetable seasonings: To control insects and microorganisms.
- Dry or dehydrated enzyme preparations: To control insects and microorganisms.
- White potatoes: To inhibit sprout development.
- Wheat and wheat flour: To control insects.
- Loose or bagged fresh iceberg lettuce and spinach.
- Crustaceans (lobster, shrimp, and crab).
- Shellfish (oysters, clams, mussels, and scallops).

Biopreservation

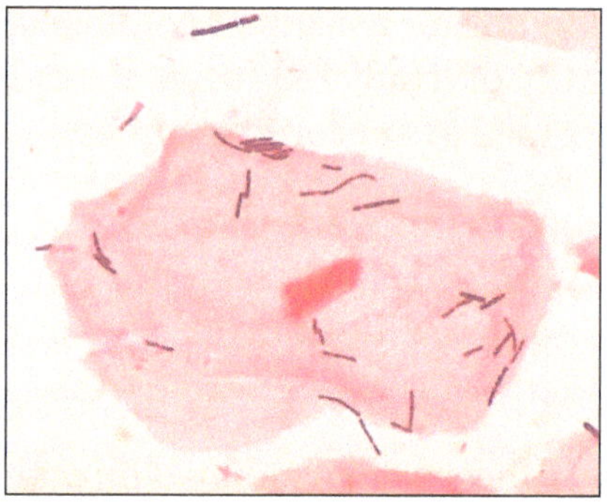

The small rods shown here are lactic acid bacteria which convert lactose and other sugars to lactic acid. The products of their metabolism can have benign preservative effects.

Biopreservation is the use of natural or controlled microbiota or antimicrobials as a way of preserving food and extending its shelf life. The biopreservation of food, especially utilizing lactic acid bacteria (LAB) that are inhibitory to food spoilage microbes, has been practiced since early ages, at first unconsciously but eventually with an increasingly robust scientific foundation. Beneficial bacteria or the fermentation products produced by these bacteria are used in biopreservation to control spoilage and render pathogens inactive in food. There are a various modes of action through which microorganisms can interfere with the growth of others such as organic acid production, resulting in a reduction of pH and the antimicrobial activity of the un-dissociated acid molecules, a wide variety of small inhibitory molecules including hydrogen peroxide, etc. It is a benign ecological approach which is gaining increasing attention.

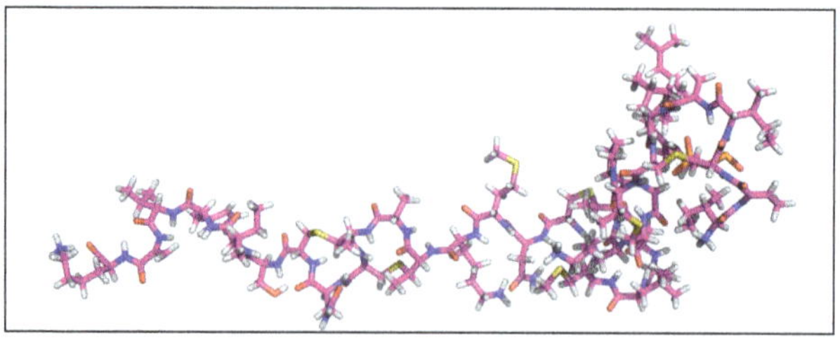

3D stick model of nisin, a particularly effective preservative produced by some lactic acid bacteria.

Biopreservative Agents and Modes of Action

Lactic Acid Bacteria

Of special interest are lactic acid bacteria (LAB). Lactic acid bacteria have antagonistic properties which make them particularly useful as biopreservatives. When LABs compete for nutrients, their metabolites often include active antimicrobials such as lactic and acetic acid, hydrogen peroxide, and peptide bacteriocins. Some LABs produce the antimicrobial nisin which is a particularly effective preservative. These days LAB bacteriocins are used as an integral part of hurdle technology. Using them in combination with other preservative techniques can effectively control spoilage bacteria and other pathogens, and can inhibit the activities of a wide spectrum of organisms, including inherently resistant Gram-negative bacteria. Lactic acid bacteria and Propionibacteria have been extensively studies for their efficacy against spoilage causing yeasts and molds in food spoilage.

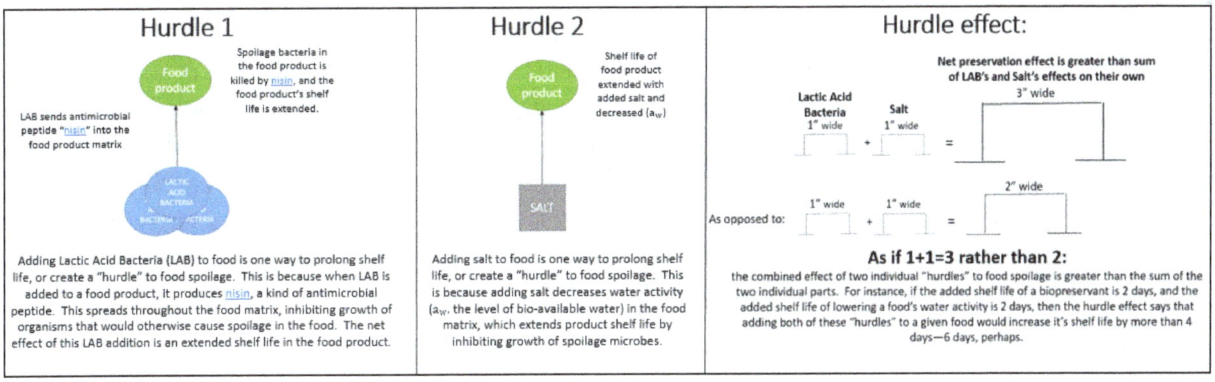

This figure illustrates the pathway of food preservation followed by lactic acid bacteria involving nisin, as well as the pathway of food preservation followed by salt. Additionally, the hurdle effect of food preservation, such as by adding lactic acid bacteria and salt to a food product, is illustrated and described.

Yeast

In addition to lactic acid bacteria, yeasts also have been reported to have a biopreservation effect due to their antagonistic activities relying on the competition for nutrients, production and tolerance of high concentrations of ethanol, as well as the synthesis of a large class of antimicrobial compounds exhibiting large spectrum of activity against food spoilage microorganisms, but also against plant, animal and human pathogen.

A bacterium/yeast that is a suitable candidate for use as a biopreservative does not necessarily have to ferment the food. However, if conditions are suitable for microbial growth, then a biopreservative bacterium will compete well for nutrients with the spoilage and pathogenic bacteria in the food. As a product of its metabolism, it should also produce acids and other antimicrobial agents, particularly bacteriocins. Biopreservative bacteria, such as lactic acid bacteria, must be harmless to humans.

Bacteriophages

Bacteriophages, or simply phages, are viruses which infect bacteria. The majority of all bacteriophages known exhibit a double-stranded DNA genome inside the virion capsid and belong to the order of tailed phages, Caudovirales. The tailed phages can be further separated into three families: Podoviridae, which are characterized by very short tails; Myoviridae, which exhibit longer, straight and contractile tails; and Siphoviridae, which can be identified due to their long and flexible tails. Another well studied group of phages with many applications, although minor in terms of species diversity, is represented by filamentous phages which exhibit a single stranded DNA genome decorated by a helical protein layer surrounding the DNA molecule. Bacteriophages are ubiquitously distributed in nature and can also be isolated from human or animal associated microflora. They outnumber their bacterial host species by a factor of ten representing the most abundant self-replicating entities on earth with an estimated 1031 phages in total. The idea of using phages against unwanted bacteria developed shortly after their discovery. With the improvements in organic chemistry during the 1950s, exploration and development of broad spectrum antibiotics displaced interest in bacteriophage research. Several laboratories have been testing suitability of bacteriophage isolates to control certain bacterial pathogens. Significant advancements in this research have been made at the Bacteriophage Institute in Tbilisi, Georgia, where phage therapy is routinely applied in medicine research field. Today treatment of antibiotic resistant bacteria is a challenging task. Recently, research on bacteriophages has gained additional momentum in light of the identification of antibiotic-resistant pathogens of infectious diseases, wherein the application of antibiotics is not effectively working, therefore research on the application of bacteriophages is being reviewed intensely. Bacteriophages have recently received a generally recognized as safe status based on their lack of toxicity and other detrimental effects to human health for application in meat products in USA.

Phage preparations specific for L. monocytogenes, E. coli O157:H7, and S. enterica serotypes have been commercialized and approved for application in foods or as part of surface decontamination protocols.

Meat Biopreservation

In meat processing, biopreservation has been extensively studied in fermented meat products and ready to eat meat products. The use of native or artificially-introduced microbial population to improve animal health and productivity, and to reduce pathogenic organisms, has been termed a probiotic or competitive enhancement approach. Competitive enhancement strategies that have been developed include competitive exclusion, addition of a microbial supplement (probiotic) that improves gastrointestinal health, and adding a limiting, non-host digestible nutrient (prebiotic) that provides an existing (or introduced) commensal microbial population a competitive advantage in the gastrointestinal tract. Each of these approaches utilizes the activities of the native microbial ecosystem against pathogens by capitalizing on the natural microbial competition. Generally speaking, competitive enhancement strategies offer a natural 'green' method to reduce pathogens in the gut of food animals.

Seafood Biopreservation

Fishery products are a source of wide variety of valuable nutrients such as proteins, vitamins, minerals, omega-3 fatty acids, taurine, etc. Fishery products, however, are also associated with human intoxication and infection. Approximately 10 to 20% of food-borne illnesses are attributed to fish consumption. Changing consumer demand has driven the appeal of traditional processes applied to seafood (e.g. salting, smoking and canning) lower compared to mild technologies involving lower salt content, lower cooking temperature and vacuum packing (VP)/modified atmosphere packing (MAP). These products, designed as lightly preserved fish products (LPFP), are usually produced from fresh seafood and further processing increases risk of cross contamination. These milder treatments are usually not sufficient to destroy microorganisms, and in some cases psychrotolerant pathogenic and spoilage bacteria can develop during the extended shelf-life of LPFP. Many of these products are also eaten raw, so minimizing the presence and preventing growth of microorganisms is essential for the food quality and safety. The microbial safety and stability of food are based on an application of preservative factors called hurdles. The delicate texture and flavor of seafood are very sensitive to the decontamination technologies such as cooking, and more recent mild technologies such as pulsed light, high pressure, ozone, and ultrasound. Chemical preservatives, which are not processes but ingredients, are out of favor with consumers due to natural preservatives demand. An alternative solution that is gaining more and more attention is biopreservation technology. In fish processing, biopreservation is achieved by adding antimicrobials or by increasing the acidity of the fish muscle. Most bacteria stop multiplying when the pH is less than 4.5. Traditionally, acidity has been increased by fermentation, marination or by directly adding acetic, citric or lactic acid to food products. Other preservatives include nitrites, sulphites, sorbates, benzoates and essential oils. The main reason for less documented studies for application of protective microorganisms, bacteriophages or bacteriocins on seafood products for biopreservation compared to dairy or meat products is probably that the early stages of biopreservation have occurred mainly in fermented foodstuffs that are not so developed among seafood products. The selection of potential protective bacteria in seafood products is challenging due to the fact that they need adaptation to the seafood matrix poor in sugar and their metabolic activities should not change the initial characteristics of the product, i.e. by acidification, and not induce spoilage that could lead to a sensory rejection. Among the microbiota identified in fresh or processed seafood, LAB remains the category that offers the highest potential for direct application as a bioprotective culture or for bacteriocin production.

Commercial Applications and Products

There has been successful implementation of various phage preparation around the globe. Various applications/delivery methods in food have been developed. Bacteriophages and their endolysins can be incorporated into food systems in several ways such as spraying, dipping or immobilization, singly or in combination with other hurdles. The phage preparation LMP-102 has been subsequently commercialized as "ListShield" Intralyx, Inc. It has been shown to be effective against 170 different strains of "L. monocytogenes", reducing significantly (10 to 1000-fold) the Listeria contamination when sprayed onto ready-to-eat foods, without changing the food general composition, taste, odor or color. The Intralytix company has also commercialized phage-based antimicrobial preparations like SalmoFresh and SalmoLyse for controlling S. enterica. SalmoFresh is prepared with a cocktail of naturally occurring lytic bacteriophages that selectively and specifically kill Salmonella, including strains belonging to the most common/highly pathogenic serotypes Typhimurium, Enteritidis, Heidelberg, Newport, Hadar, Kentucky and Thompson. According to the manufacturer, SalmoFresh is specifically designed for treating foods that are at high risk for "Salmonella" contamination. In particular, red meat and poultry can be treated prior to grinding for significant reductions in Salmonella contamination. SalmoLyse is a reformulated phage cocktail derived from SalmoFresh in which two of the six phages in the original cocktail have been replaced. Additional bacteriophage preparations have been formulated and referenced to be used to reduce the microbial load of animals prior to slaughter and are commercially available from

Omnilytics such as the BacWash product line against Salmonella Omnilytics. Another commercial application has been developed, Listex_ P100 by Micreos in The Netherlands and was granted generally recognized as safe (GRAS) status by the FDA and USDA for use in all food products.

Another significant commercial bacteriophage application is ELICOSALI, a wide range of anti-Salmonella and "E. coli" phage cocktail, for treatment of agricultural products developed by Eliava Institute at Tbilisi, Republic of Georgia Eliava Institute.

Safety

Biopreservation judiciously exploits the antimicrobial potential of naturally occurring microorganisms in food and their metabolites with a long history of safe use. Bacteriocins, bacteriophages and bacteriophage-encoded enzymes fall in this theory. The long and traditional role of Lactic acid bacteria on food and feed fermentations is the main factor related to the use of bacteriocins in biopreservation. LAB and their bacteriocins have been consumed unintentionally for ages, laying down a long history of safe use. Their antimicrobial spectrum of inhibition, bactericidal mode of action, relative tolerance to processing conditions (pH, NaCl, heat treatments) and the lack of toxicity towards eukaryotic cells enforces their role as biopreservatives in food. The evaluation of any new antimicrobial actives is done in meat by USDA which relies on the GRAS assessment by FDA among other suitability data.

Vacuum Packing

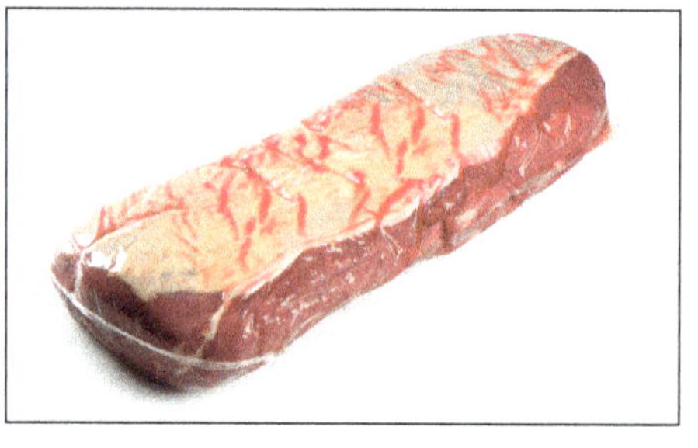

Vacuum packaging is another way to increase the shelf life of food products. Here the product is placed in an air-tight pack, the air sucked out and the package sealed. By removing air from around the product, the levels of oxygen in the packaging are reduced, impeding the ability of oxygen-breathing microorganisms to grow and spoil the product. The lack of oxygen also reduces the amount of spoilage due to oxidation – the process that causes apples and bananas to turn brown, for example.

A certain amount of oxygen will remain, however, because it is not possible to create a total vacuum. Air contains around 21 per cent oxygen at normal atmospheric pressure – 1000 millibar. As the air is withdrawn during the vacuum packaging process, the pressure inside the package is reduced.

If for example, the pressure is reduced to 100 millibar, an equivalent of around 2.1 per cent oxygen will remain; if it is reduced to 10 millibar, there will still be in effect 0.21 per cent oxygen present.

Is vacuum packaging more effective than modified atmosphere packaging? As with most things, it is a case of 'horses for courses' – it depends on the product being packaged.

MAP is certainly a more versatile process than vacuum packaging. Vacuum packaging is essentially a 'one size fits all' technology – it relies solely on removing air. MAP on the other hand can be tailored to the particular foodstuff, with different gases and different proportions of gas in the mixture used to give the maximum shelf life for the particular product and to retain the quality and appearance of the product.

For example for packaging seafood, the proportion of carbon dioxide and oxygen in the modified atmosphere can by varied depending on the type of fish: oily fish benefit from a different atmosphere compared with, for example, prawns. The gas mixture used for red meat is different to that used for, say, bread.

One area where MAP does score well when compared with vacuum packaging is in the presentation of the product. In vacuum packaging, as the pressure within the packaging is reduced the packaging material collapses and forms itself tightly around the product. For some products, such as fresh meat, this can distort the appearance of the product. Other products, such as shredded cheese, are not suitable for vacuum packaging because the pressure of the packaging material on the product would cause the product to deform and to lose important characteristics.

Another aspect in which the two processes differ is in the ease of quality control of the packaging process. In MAP packaging, air is flushed from the package and replaced with the gas mixture, making it possible to constantly monitor the gas content of the package during the packaging process. Once the package is sealed, any leakage of the modified atmosphere can be detected, enabling the integrity of the seal to be assured. For vacuum packaging, because there is no gas present in the package, leak testing is typically done through manual inspection, making quality control less straightforward.

References

- Vanlinit, D.; et al. (2011). "Rapid Acquisition of Gigapascal-High-Pressure Resistance by Escherichia coli". Mbio. 2 (1): e00130–10. Doi:10.1128/mbio.00130-10. PMC 3025523. PMID 21264062
- Ohlsson T and Bengtsson N (2002) "The hurdle concept" Minimal Processing Technologies in the Food Industry, Woodhead Publishing, pp. 175–195. ISBN 978-1-85573-547-7
- Pasteurization, biochemistry, science-and-technology: encyclopedia.com, Retrieved 19 April, 2019
- Balasubramaniam, V.M., Barbosa-Cánovas, Gustavo V., Lelieveld, Huub L.M (2016). High Pressure Processing of Food Principles, Technology and Applications. Springer. ISBN 978-1-4939-3234-4
- Fan Xuetong (2011). "Changes in Quality, Liking, and Purchase Intent of Irradiated Fresh-Cut Spinach during Storage". Journal of Food Science (Submitted manuscript). 76 (6): S363–S368. Doi:10.1111/j.1750-3841.2011.02207.x. PMID 21623783
- FAO: Preservation techniques Fisheries and aquaculture department, Rome. Updated 27 May 2005. Retrieved 14 March 2011
- Alzamora, Stella; Tapia, Maria Soledad; López-Malo, Aurelio (2000). Minimally Processed Fruits and Vegetables: Fundamental aspects and applications. Springer. P. 266. ISBN 978-0-8342-1672-3
- Vacuum-packaging, modified-atmosphere-packaging-resources: modifiedatmospherepackaging.com, Retrieved 5 February, 2019

4
Preservation Techniques

The food can be preserved by using many preservation techniques such as salting, pickling, sugaring, canning, food drying, smoking, refrigeration and by using fruit preserves, jams and jellies. This chapter closely examines these key preservation techniques to provide an extensive understanding of the subject.

Salting Food

Salt has been used as a preservative since ancient times and was, in fact, so valuable in part due to its preservative powers that it became one of the first ever commodities traded and even wars have been fought over salt.

Before the days of refrigeration and the modern heat treatments that are used to prolong the shelf life of foods, salt was one of the main methods used to preserve food. Without it people were subject to having to consume food quickly before it would spoil, with salt civilisations were able to store food and build up reserves that would last in times of scarcity.

The Simple Reason why Salt Works to Preserve Food

The reason why salt is such an effective preservative is that it draws moisture out of food. Dry foods are less likely to spoil as moisture is a key requirement for organisms that want to spoil the food.

When something like meat is exposed to salt in the right quantity, around 20% salinity, the salt begins to draw the moisture out of the cells in not only the food but also bacteria present in the food and this makes it last longer because the bacteria that will spoil the meat have a hard time surviving in such a salty environment with little moisture.

How does Salt Prevent Spoilage?

Spoilage organisms, such as bacteria, yeasts and mould grow on the food making it undesirable or just unsafe to eat all require moisture to feed on the food. The moisture content of the food is, therefore, a critical aspect of how likely the food it to spoil in a short time frame.

We know then that by reducing the moisture content of foods we can inhibit microorganisms from growing and therefore extend the shelf life of the food.

Dehydrating works on this same principle, many types of food such as fruits, vegetables, meat and fish is dried. Ancient ancestors in the right climates would have dried meat in the sun to save food from a hunt. In areas where the climate is not suitable to dry foods in the sun, our ancestors would have has to come up with a different method of dehydrating the food, salt was one of these ways.

The mechanism that is in effect when we salt foods to preserve them is osmosis. Osmosis works by drawing water across a cell membrane so that both sides have an equal amount of saltiness or salinity.

Salt not only draws water out of food but also draws moisture out of any bacteria or microorganism that is on or in the food killing them in the process.

Brine or Dry Salting

When we cure foods with salt it can either be done by applying actual salt granules to the food or via a brine that the food is immersed in. Both processes draw moisture out of the food to dehydrate it but some applications work in slightly different ways that affect how the food is preserved.

Salted Fish

Cod is often preserved in salt using a 20% brine solution. At this level of 20% salinity, water is effectively drawn out of any microorganism that will spoil the meat. The fish will last several months in the fridge longer than it would without the salt. The same process can also be done with coarse salt without the brine. Fillets of cod is packed in coarse salt and sold commercially as "salt cod".

If we pack food with salt, the concentration of salt causes an imbalance. Osmosis works to try and balance the salinity of the cells of the food with the salinity around the food. As salt is drawn into the food water is drawn out. As the water is drawn out the food gets drier and drier making it an inhospitable place for food spoilage organisms to grow.

At this level of salinity, however, the fish will be extremely salty if eaten as is. The fish will need to be soaked in water for at least 24 – 48 hours in freshwater to draw out a lot of the salt from the fish.

Salt used in Fermentation

Vegetables are regularly preserved in brine but do not rely on just salt as a preservative. Dill Pickles are packed in jars win a brine of about 3 – 5 % salinity. At this level, the salt isn't strong enough as a preservative on its own. Most bacteria cannot survive but some can, fortunately, friendly bacteria called lactobacillus are one of a few types of bacteria that thrive in brine like this. The lactobacillus ferments the sugars present in the vegetables and turns the brine acidic. The salty and acidic environment pickles the vegetables and stops any food spoilage organisms from growing.

A similar process is used to make Sauerkraut, a pickled cabbage, except instead of immersing the cabbage in brine, dry salt is sprinkled over the cabbage and then packed into jars. The salt prevents most bacteria from growing except for lactobacillus that then creates an acidic environment and the combined preservative effect of salt and acidity preserve the sauerkraut for months even without refrigeration.

Curing Salts – Nitrates and Nitrites

Meat is often cured with salt with a whole subsection of foods preserved meats called charcuterie. Curing meat with salt relies on the same principles of drawing moisture out of the meat and spoilage organisms.

However, salt alone is not safe enough for moderns food safety and hygiene. Another type of salt called a curing salt is used. This curing salt is often referred to as Prague powder or instacure.

The addition of nitrates or nitrites to salt is far more effective at inhibiting bacterial growth in preserved meats.

Prague powder or curing salt is a mixture of regular salt and a compound called sodium nitrate or sodium nitrite depending on the type of cure being done. The salt is pink to differentiate it to regular table salt.

Curing salts with the addition of sodium nitrate or sodium nitrate are carefully measured according to the amount of meat being cured. They enhance both the flavour and colour of the meat as well as enhancing the unique flavour that is synonymous with cured meat.

Salt as a Preservative

Salt has many culinary uses and as a preservative has a long history in a variety of applications. It is used to pickle vegetable, dry and cures meat or fish and at the same time greatly enhances and changes the texture and flavour of them.

This all works on the basic principle of osmosis drawing water from cells whilst drawing salt into the food to create balance.

Curing

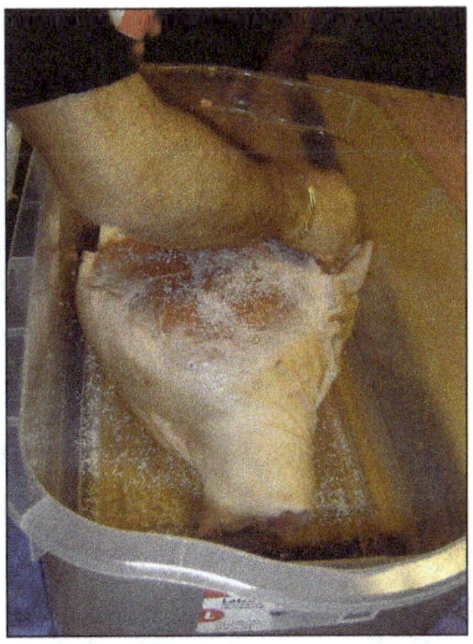

Sea salt being added to raw ham to make prosciutto.

Curing is any of various food preservation and flavoring processes of foods such as meat, fish and vegetables, by the addition of salt, with the aim of drawing moisture out of the food by the process of osmosis. Because curing increases the solute concentration in the food and hence decreases its water potential, the food becomes inhospitable for the microbe growth that causes food spoilage. Curing can be traced back to antiquity, and was the primary way of preserving meat and fish until the late-19th century. Dehydration was the earliest form of food curing. Many curing processes also involve smoking, spicing, cooking, or the addition of combinations of sugar, nitrate, nitrite.

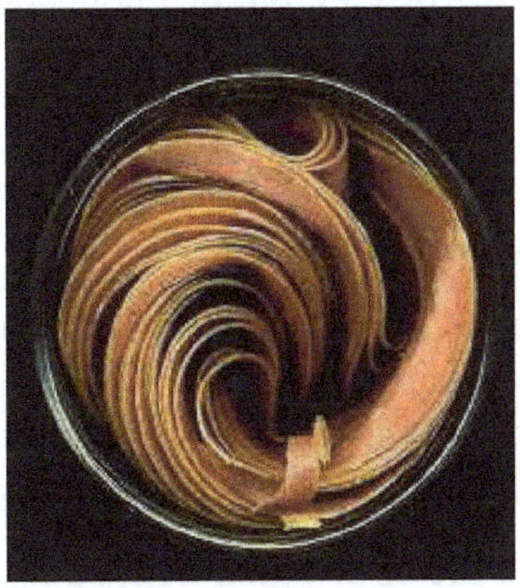

Slices of beef in a can.

Meat preservation in general (of meat from livestock, game, and poultry) comprises the set of all treatment processes for preserving the properties, taste, texture, and color of raw, partially cooked, or cooked meats while keeping them edible and safe to consume. Curing has been the dominant method of meat preservation for thousands of years, although modern developments like refrigeration and synthetic preservatives have begun to complement and supplant it.

While meat-preservation processes like curing were mainly developed in order to prevent disease and to increase food security, the advent of modern preservation methods mean that in most developed countries today curing is instead mainly practised for its cultural value and desirable impact on the texture and taste of food. For lesser-developed countries, curing remains a key process in the production, transport and availability of meat.

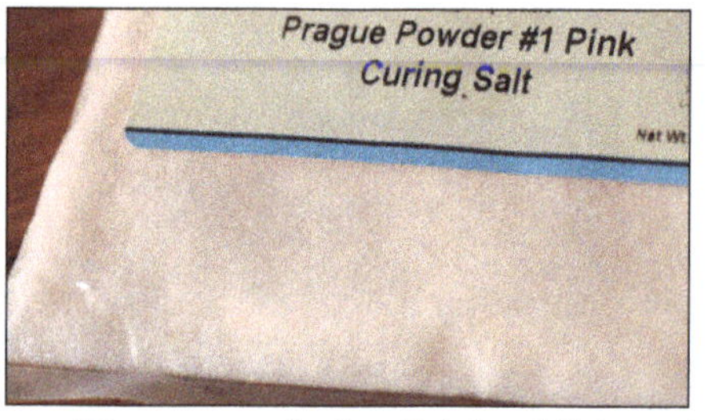

Curing salt, also known as "Prague powder" or "pink salt", is typically a combination of sodium chloride and sodium nitrite that is dyed pink to distinguish it from table salt.

Some traditional cured meat (such as authentic Parma ham and some authentic Spanish chorizo and Italian salami) are cured with salt alone. Today, potassium nitrate and sodium nitrite (in conjunction with salt) are the most common agents in curing meat, because they bond to the myoglobin and act as a substitute for the oxygen, thus turning myoglobin red. More recent evidence shows that these chemicals also inhibit the growth of the bacteria that cause the disease botulism. The combination of table salt with nitrates or nitrites, called curing salt, is often dyed pink to distinguish it from table salt. Neither table salt, nor any of the nitrites or nitrates commonly used in curing (e.g. sodium nitrate, sodium nitrite, and potassium nitrate) is naturally pink.

Necessity of Curing

Untreated meat decomposes rapidly if it is not preserved, at a speed that depends on several factors, including ambient humidity, temperature, and the presence of pathogens. Most meats cannot be kept at room temperature in excess of a few days without spoiling.

If kept in excess of this time, meat begins to change color and exude a foul odor, indicating the decomposition of the food. Ingestion of such spoiled meat can cause serious food poisonings, like botulism. Salt-curing processes have been developed since antiquity in order to ensure food safety without relying on artificial anti-bacterial agents.

While the short shelf life of fresh meat does not pose a significant problem when access to it is easy and supply is abundant, in times of scarcity and famine, or when the meat must be carried over long voyages, it spoils very quickly. In such circumstances the usefulness of preserving foods containing nutritional value for transport and storage is obvious.

Curing can significantly extend the life of meat before it spoils, by making it inhospitable to the growth of spoilage microbes.

Chemical Actions

Salt

Salt (sodium chloride) is the primary ingredient used in meat curing. Removal of water and addition of salt to meat creates a solute-rich environment where osmotic pressure draws water out of microorganisms, slowing down their growth. Doing this requires a concentration of salt of nearly 20%. In addition, salt causes the soluble proteins to come to the surface of the meat that was used to make the sausages. These proteins coagulate when the sausage is heated, helping to hold the sausage together.

Sugar

The sugar added to meat for the purpose of curing it comes in many forms, including honey, corn syrup solids, and maple syrup. However, with the exception of bacon, it does not contribute much to the flavor, but it does alleviate the harsh flavor of the salt. Sugar also contributes to the growth of beneficial bacteria such as *Lactobacillus* by feeding them.

Nitrates and Nitrites

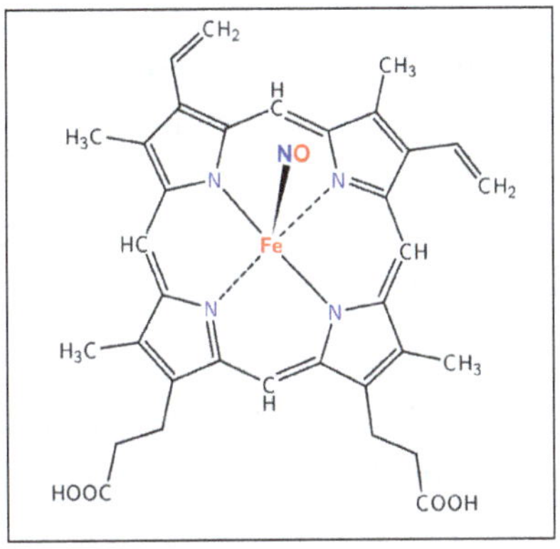

Nitrosyl-heme.

Nitrates and nitrites not only help kill bacteria, but also produce a characteristic flavor and give meat a pink or red color. Nitrite (NO^-_2) is generally supplied by sodium nitrite or (indirectly) by potassium nitrate. Nitrite salts are most often used in curing. Nitrate is specifically used only in a few curing conditions and products where nitrite (which may be generated from nitrate) must be generated in the product over long periods of time.

Nitrite further breaks down in the meat into nitric oxide (NO), which then binds to the iron atom in the center of myoglobin's heme group, reducing oxidation and causing a reddish-brown color (nitrosomyoglobin) when raw and the characteristic cooked-ham pink color (nitrosohemochrome or nitrosyl-heme) when cooked. The addition of ascorbate to cured meat reduces formation of nitrosamines, but increases the nitrosylation of iron.

The use of nitrite and nitrate salts for meat in the US has been formally used since 1925. Because of the relatively high toxicity of nitrite (the lethal dose in humans is about 22 mg/kg of body weight), the maximum allowed nitrite concentration in meat products is 200 ppm. Plasma nitrite is reduced in persons with endothelial dysfunction.

The use of nitrites in food preservation is controversial due to the potential for the formation of nitrosamines when nitrites are present in high concentrations and the product is cooked at high temperatures. The effect is seen for red or processed meat, but not for white meat or fish. Nitrates and nitrites may cause cancer and the production of carcinogenic nitrosamines can be potently inhibited by the use of the antioxidants Vitamin C and the alpha-tocopherol form of Vitamin E during curing. Under simulated gastric conditions, nitrosothiols rather than nitrosamines are the main nitroso species being formed. The use of either compound is therefore regulated; for example, in the United States, the concentration of nitrates and nitrites is generally limited to 200 ppm or lower. While the meat industry considers them irreplaceable because of their low cost and efficacy at maintaining color, botulism is an extremely rare disease (less than 1000 cases reported worldwide per year), and almost always associated with home preparations of food storing. Furthermore, while the FDA has set a limit of 200 ppm of nitrates for cured meat, they are not allowed and not recognized as safe in most other foods, even foods that are not cooked at high temperatures, such as cheese.

Nitrites from Celery

Processed meats without "added nitrites" may be misleading as they may be using naturally occurring nitrites from celery instead.

A 2019 report from Consumer Reports found that using celery (or other natural sources) as a curing agent introduced naturally occurring nitrates and nitrites. The USDA allows the term "uncured" or "no nitrates or nitrites added" on products using these natural sources of nitrites, which provides the consumer a false sense of making a healthier choice. The Consumer Reports investigation also provides the average level of sodium, nitrates and nitrites found per gram of meat in their report.

Consumer Reports and the Center for Science in the Public Interest filed a formal request to the USDA to change the labeling requirements this year.

Smoke

Meat can also be preserved by "smoking". If the smoke is hot enough to slow-cook the meat, this will also keep it tender. One method of smoking calls for a smokehouse with damp wood chips or sawdust. In North America, hardwoods such as hickory, mesquite, and maple are commonly used for smoking, as are the wood from fruit trees such as apple, cherry, and plum, and even corncobs.

Smoking helps seal the outer layer of the food being cured, making it more difficult for bacteria to enter. It can be done in combination with other curing methods such as salting. Common smoking styles include hot smoking, smoke roasting (pit barbecuing) and cold smoking. Smoke roasting and hot smoking cook the meat while cold smoking does not. If the meat is cold smoked, it should be dried quickly to limit bacterial growth during the critical period where the meat is not yet dry. This can be achieved, as with jerky, by slicing the meat thinly.

The smoking of food directly with wood smoke is known to contaminate the food with carcinogenic polycyclic aromatic hydrocarbons.

Effect of Meat Preservation

On Health

Since the 20th century, with respect to the relationship between diet and human disease (e.g. cardiovascular, etc.), scientists have conducted studies on the effects of lipolysis on vacuum-packed or frozen meat. In particular, by analyzing entrecôtes of frozen beef during 270 days at −20 °C (−4 °F), scientists found an important phospholipase that accompanies the loss of some unsaturated fat n-3 and n-6, which are already low in the flesh of ruminants.

In 2015, the International Agency for Research on Cancer of the World Health Organization classified processed meat, that is, meat that has undergone salting, curing, fermenting, or smoking, as "carcinogenic to humans".

On Trade

The improvement of methods of meat preservation, and of the means of transport of preserved products, has notably permitted the separation of areas of production and areas of consumption, which can now be distant without it posing a problem, permitting the exportation of meats.

For example, the appearance in the 1980s of preservation techniques under controlled atmosphere sparked a small revolution in the world's market for sheep meat: the lamb of New Zealand, one of the world's largest exporters of lamb, could henceforth be sold as fresh meat, since it could be preserved from 12 to 16 weeks, which would be a sufficient duration for it to reach Europe by boat. Before, meat from New Zealand was frozen, thus had a much lower value on European shelves. With the arrival of the new "chilled" meats, New Zealand could compete even more strongly with local producers of fresh meat. The use of controlled atmosphere to avoid the depreciation which affects frozen meat is equally useful in other meat markets, such as that for pork, which now also enjoys an international trade.

Pickling

A jar of pickled cucumbers (front) and a jar of pickled onions (back).

Pickling is the process of preserving or extending the lifespan of food by either anaerobic fermentation in brine or immersion in vinegar. In East Asia, vinaigrette (vegetable oil and vinegar) is also used as a pickling medium. The pickling procedure typically affects the food's texture, taste and flavor. The resulting food is called a *pickle*, or, to prevent ambiguity, prefaced with *pickled*. Foods that are pickled include vegetables, fruits, meats, fish, dairy and eggs.

A distinguishing characteristic is a pH of 4.6 or lower, which is sufficient to kill most bacteria. Pickling can preserve perishable foods for months. Antimicrobial herbs and spices, such as mustard seed, garlic, cinnamon or cloves, are often added. If the food contains sufficient moisture, a pickling brine may be produced simply by adding dry salt. For example, sauerkraut and Korean kimchi are produced by salting the vegetables to draw out excess water. Natural fermentation at room temperature, by lactic acid bacteria, produces the required acidity. Other pickles are made by placing vegetables in vinegar. Like the canning process, pickling (which includes fermentation) does not require that the food be completely sterile before it is sealed. The acidity or salinity of the solution, the temperature of fermentation, and the exclusion of oxygen determine which microorganisms dominate, and determine the flavor of the end product.

When both salt concentration and temperature are low, *Leuconostoc mesenteroides* dominates, producing a mix of acids, alcohol, and aroma compounds. At higher temperatures *Lactobacillus plantarum* dominates, which produces primarily lactic acid. Many pickles start with *Leuconostoc*, and change to *Lactobacillus* with higher acidity.

Europe

Central and Eastern Europe

Coriander seeds are one of the spices popularly added to pickled vegetables in Europe.

In Hungary the main meal (lunch) usually goes with some kind of pickles but they are commonly consumed at other times of the day too. The most commonly consumed pickles are sauerkraut, the different kinds of pickled cucumbers and peppers and csalamádé but tomatoes, carrots, beetroot, baby corn, onions, garlic, certain squashes and melons and a few fruits like plums and apples are used to make pickles too. Stuffed pickles are specialties usually made of peppers or melons pickled after being stuffed with a cabbage filling. Pickled plum stuffed with garlic is a unique Hungarian type of pickle just like csalamádé and leavened cucumber. Csalamádé a type of mixed pickle made of cabbage, cucumber, paprika, onion, carrot, tomatoes and bay leaf mixed up with vinegar as the fermenting agent. Leavened cucumber, unlike other types of pickled cucumbers that are around all year long, is rather a seasonal pickle produced in the summer. Cucumbers, spices, herbs and slices of bread are put in a glass jar with salt water and kept in direct sunlight for a few days. The yeast from the bread, along with other pickling agents and spices fermented under the hot sun, give the cucumbers a unique flavor, texture and slight carbonation. Its juice can be used to make a special type of spritzer instead of carbonated water. It is common for Hungarian households to produce their own pickles. Different regions or towns have their special recipes unique to them. Among them all the Vecsési sauerkraut is the most famous.

Jonjoli Georgian pickled flowers of bladdernut.

Pickled tomatoes are common in CIS.

Romanian pickles are made out of beetroot, cucumbers, green tomatoes (*gogonele*), carrots, cabbage, garlic, sauerkraut (bell peppers stuffed with cabbage), bell peppers, melons, mushrooms, turnips, celery and cauliflower. Meat, like pork, can also be preserved in salt and lard.

Polish, Czech and Slovak traditional pickles are cucumbers and sauerkraut, but other pickled fruits and vegetables, including plums, pumpkins and mushrooms are also common.

Russian, Ukrainian and Belarusian pickled items include beets, mushrooms, tomatoes, sauerkraut, cucumbers, ramsons, garlic, eggplant (which is typically stuffed with julienned carrots), custard squash, and watermelon. Garden produce is commonly pickled using salt, dill, blackcurrant leaves, bay leaves and garlic and is stored in a cool, dark place. The leftover brine (called rassol (рассол) in Russian) has a number of culinary uses in these countries, especially for cooking traditional soups, such as shchi, rassolnik, and solyanka. Rassol, especially cucumber or sauerkraut rassol, is also a favorite traditional remedy against morning hangover.

Southern Europe

An Italian pickled vegetable dish is giardiniera, which includes onions, carrots, celery and cauliflower. Many places in southern Italy, particularly in Sicily, pickle eggplants and hot peppers.

In Albania, Bulgaria, Serbia, Macedonia and Turkey, mixed pickles, known as turshi, tursija or turshu form popular appetizers, which are typically eaten with rakia. Pickled green tomatoes, cucumbers, carrots, bell peppers, peppers, eggplants, and sauerkraut are also popular.

Turkish pickles, called tursu, are made out of vegetables, roots, and fruits such as peppers, cucumber, Armenian cucumber, cabbage, tomato, eggplant (aubergine), carrot, turnip, beetroot, green almond, baby watermelon, baby cantaloupe, garlic, cauliflower, bean and green plum. A mixture of spices flavor the pickles.

Northern Europe

In Britain, pickled onions and pickled eggs are often sold in pubs and fish and chip shops. Pickled beetroot, walnuts, and gherkins, and condiments such as Branston Pickle and piccalilli are typically eaten as an accompaniment to pork pies and cold meats, sandwiches or a ploughman's lunch. Other popular pickles in the UK are pickled mussels, cockles, red cabbage, mango chutney, sauerkraut, and olives. Rollmops are also quite widely available under a range of names from various producers both within and out of the UK.

Pickled herring, rollmops, and salmon are popular in Scandinavia. Pickled cucumbers and red garden beets are important as condiments for several traditional dishes. Pickled capers are also common in Scandinavian cuisine.

United States and Canada

A dish of giardiniera.

In the United States and Canada, pickled cucumbers (most often referred to simply as "pickles"), olives, and sauerkraut are most commonly seen, although pickles common in other nations are also available. In Canada, there may be a distinction made between gherkins (usually smaller), and pickles (larger pickled cucumbers).

Canadian pickling is similar to that of Britain. Through the winter, pickling is an important method of food preservation. Pickled cucumbers, onions, and eggs are common individual pickled foods seen in Canada. Pickled egg and pickled sausage make popular pub snacks in much of English Canada. Chow-chow is a tart vegetable mix popular in the Maritime Provinces and the Southern United States, similar to piccalilli. Pickled fish is commonly seen, as in Scotland, and kippers may be seen for breakfast, as well as plentiful smoked salmon. Meat is often also pickled or preserved in different brines throughout the winter, most prominently in the harsh climate of Newfoundland.

In the United States, Giardiniera, a mixture of pickled peppers, celery and olives, is a popular condiment in Chicago and other cities with large Italian-American populations, and is often consumed with Italian beef sandwiches. Pickled eggs are common in the Upper Peninsula of Michigan. Pickled herring is available in the Upper Midwest. Pennsylvania Dutch Country has a strong tradition of pickled foods, including chow-chow and red beet eggs. In the Southern United States, pickled okra and watermelon rind are popular, as are deep-fried pickles and pickled pig's feet, pickled chicken eggs, pickled quail eggs, pickled garden vegetables and pickled sausage. In Mexico, chili peppers, particularly of the Jalapeño and serrano varieties, pickled with onions, carrots and herbs form common condiments. Various pickled vegetables, fish, or eggs may make a side dish to a Canadian lunch or dinner. Popular pickles in the Pacific Northwest include pickled asparagus and green beans. Pickled fruits like blueberries and early green strawberries are paired with meat dishes in restaurants.

In some parts of the United States, pickles with Kool-Aid are a popular food for children. In the United States, National Pickle Day is recognized as a food "holiday" every year on November 14.

Mexico, Central America and South America

In the Mesoamerican region pickling is known as "encurtido" or "curtido" for short. The pickles or "curtidos" as known in Latin America are served cold, as an appetizer, as a side dish or as a tapas dish in Spain. In several Central American countries it is prepared with cabbage, onions, carrots, lemon, vinegar, oregano, and salt. In Mexico, "curtido" consists of carrots, onions, and jalapeño peppers and used to accompany meals still common in taquerias and restaurants. In order to prepare a carrot "curtido" simply add carrots to vinegar and other ingredients that are common to the region such as chilli, tomato, and onions. Varies depending on the food, in the case of sour. Another example of a type of pickling which involves the pickling of meats or seafood is the "escabeche" or "ceviches" popular in Peru, Ecuador, and throughout Latin America and the Caribbean. These dishes include the pickling of pig's feet, pig's ears, and gizzards prepared as an "escabeche" with spices and seasonings to flavor it. The ceviches consists of shrimp, octopus, and various fishes seasoned and served cold.

Process

Bat Trang porcelain vessel for pickling.

In traditional pickling, fruit or vegetables are submerged in brine (20-40 grams/L of salt (3.2–6.4 oz/imp gal or 2.7–5.3 oz/US gal)), or shredded and salted as in sauerkraut preparation, and held underwater by flat stones layered on top. Alternatively, a lid with an airtrap or a tight lid may be used if the lid is able to release pressure which may result from carbon dioxide buildup. Mold or (white) kahm yeast may form on the surface; kahm yeast is mostly harmless but can impart an off taste and may be removed without affecting the pickling process.

In chemical pickling, the fruits or vegetables to be pickled are placed in a sterilized jar along with brine, vinegar, or both, as well as spices, and are then allowed to mature until the desired taste is obtained.

The food can be pre-soaked in brine before transferring to vinegar. This reduces the water content of the food, which would otherwise dilute the vinegar. This method is particularly useful for fruit and vegetables with a high natural water content.

In commercial pickling, a preservative such as sodium benzoate or EDTA may also be added to enhance shelf life. In fermentation pickling, the food itself produces the preservation agent, typically by a process involving Lactobacillus bacteria that produce lactic acid as the preservative agent.

Alum is used in pickling to promote crisp texture and is approved as a food additive by the United States Food and Drug Administration.

"Refrigerator pickles" are unfermented pickles made by marinating fruit or vegetables in a seasoned vinegar solution. They must be stored under refrigeration or undergo canning to achieve long-term storage.

Japanese Tsukemono use a variety of pickling ingredients depending on their type, and are produced by combining these ingredients with the vegetables to be preserved and putting the mixture under pressure.

Possible Health Hazards of Pickled Vegetables

The World Health Organization has listed pickled vegetables as a possible carcinogen, and the British Journal of Cancer released an online 2009 meta-analysis of research on pickles as increasing the risks of esophageal cancer. The report, citing limited data in a statistical meta analysis, indicates a potential two-fold increased risk of oesophageal cancer associated with Asian pickled vegetable consumption. Results from the research are described as having "high heterogeneity" and the study said that further well-designed prospective studies were warranted. However, their results stated "The majority of subgroup analyses showed a statistically significant association between consuming pickled vegetables and Oesophageal Squamous Cell Carcinoma".

The 2009 meta-analysis reported heavy infestation of pickled vegetables with fungi. Some common fungi can facilitate the formation of N-nitroso compounds, which are strong oesophageal carcinogens in several animal models. Roussin red methyl ester, a non-alkylating nitroso compound with tumour-promoting effect in vitro, was identified in pickles from Linxian in much higher concentrations than in samples from low-incidence areas. Fumonisin mycotoxins have been shown to cause liver and kidney tumours in rodents.

A 2017 study in Chinese Journal of Cancer has linked salted vegetables (common among Chinese cuisine) to a fourfold increase in nasopharynx cancer, where fermentation was a critical step in creating nitrosamines, which some are confirmed carcinogens, as well as activation of Epstein–Barr virus by fermentation products.

Historically, pickling caused health concerns for reasons associated with copper salts, as explained in the mid-19th century The English and Australian Cookery Book: "The evidence of the Lancet commissioner (Dr. Hassall) and Mr. Blackwell (of the eminent firm of Crosse and Blackwell) went to prove that the pickles sold in the shops are nearly always artificially coloured, and are thus rendered highly unwholesome, if not actually poisonous."

Refrigeration

Commercial refrigeration.

Refrigeration is the process of cooling a space, substance, or system to lower and maintain its temperature below the ambient one (while the removed heat is rejected at a higher temperature).

In other words, refrigeration means artificial (human-made) cooling. Heat is removed from a low-temperature reservoir and transferred to a high-temperature reservoir. The work of heat transfer is traditionally driven by mechanical means, but can also be driven by heat, magnetism, electricity, laser, or other means. Refrigeration has many applications, including, but not limited to: household refrigerators, industrial freezers, cryogenics, and air conditioning. Heat pumps may use the heat output of the refrigeration process, and also may be designed to be reversible, but are otherwise similar to air conditioning units.

Refrigeration has had a large impact on industry, lifestyle, agriculture, and settlement patterns. The idea of preserving food dates back to at least the ancient Roman and Chinese empires. However, mechanical refrigeration technology has rapidly evolved in the last century, from ice harvesting to temperature-controlled rail cars. The introduction of refrigerated rail cars contributed to the westward expansion of the United States, allowing settlement in areas that were not on main transport channels such as rivers, harbors, or valley trails. Settlements were also developing in infertile parts of the country, filled with newly discovered natural resources.

These new settlement patterns sparked the building of large cities which are able to thrive in areas that were otherwise thought to be inhospitable, such as Houston, Texas, and Las Vegas, Nevada. In most developed countries, cities are heavily dependent upon refrigeration in supermarkets, in order to obtain their food for daily consumption. The increase in food sources has led to a larger concentration of agricultural sales coming from a smaller percentage of existing farms. Farms today have a much larger output per person in comparison to the late 1800s. This has resulted in new food sources available to entire populations, which has had a large impact on the nutrition of society.

As quite similar criteria shall be fulfilled by working fluids (refrigerants) applied to heat pumps, refrigeration, and organic Rankine cycles; several working fluids are applied by all these technologies. Ammonia was one of the first refrigerants. Refrigeration can be defined as "The science of providing and maintaining temperature below that of surrounding atmosphere". It means continuous extraction of heat from a body whose temperature is already below the temperature of its surroundings.

Impact on Settlement Patterns

In the last century refrigeration allowed new settlement patterns to emerge. This new technology has allowed for new areas to be settled that are not on a natural channel of transport such as a river, valley trail or harbor that may have otherwise not been settled. Refrigeration has given opportunities to early settlers to expand westward and into rural areas that were unpopulated. These new settlers with rich and untapped soil saw opportunity to profit by sending raw goods to the eastern cities and states. In the 20th century, refrigeration has made "Galactic Cities" such as Dallas, Phoenix and Los Angeles possible.

Refrigerated Rail Cars

The refrigerated rail car (refrigerated van or refrigerator car), along with the dense railroad network, became an exceedingly important link between the marketplace and the farm allowing for a national opportunity rather than a just a regional one. Before the invention of the refrigerated rail car it was impossible to ship perishable food products long distances. The beef packing industry made the first demand push for refrigeration cars. The railroad companies were slow to adopt this new invention because of their heavy investments in cattle cars, stockyards, and feedlots. Refrigeration cars were also complex and costly compared to other rail cars, which also slowed the adoption of the refrigerated rail car. After the slow adoption of the refrigerated car, the beef packing industry dominated the refrigerated rail car business with their ability to control ice plants and the setting of icing fees. The United States Department of Agriculture estimated that in 1916 over sixty-nine percent of the cattle killed in the country was done in plants involved in interstate trade. The same companies that were also involved in the meat trade later implemented refrigerated transport to include vegetables and fruit. The meat packing companies had much of the expensive machinery, such as refrigerated cars, and cold storage facilities that allowed for them to effectively distribute all types of perishable goods. During World War I, a national refrigerator car pool was established by the United States Administration to deal with problem of idle cars and was later continued after the war. The idle car problem was the problem of refrigeration cars sitting pointlessly in between seasonal harvests. This meant that very expensive cars sat in rail yards for a good portion of the year while making no revenue for the car's owner. The car pool was a system where cars were distributed to areas as crops matured ensuring maximum use of the cars. Refrigerated rail cars moved eastward from vineyards, orchards, fields, and gardens in western states to satisfy Americas consuming market in the east. The refrigerated car made it possible to transport perishable crops hundreds and even thousands of miles. The most noticeable effect the car gave was a regional specialization of vegetables and fruits. The refrigeration rail car was widely used for the transportation of perishable goods up until the 1950s. By the 1960s the nation's interstate highway system was adequately complete allowing for trucks to carry the majority of the perishable food loads and to push out the old system of the refrigerated rail cars.

Expansion West and into Rural Areas

The widespread use of refrigeration allowed for a vast amount of new agricultural opportunities to open up in the United States. New markets emerged throughout the United States in areas that were previously uninhabited and far-removed from heavily populated areas. New agricultural opportunity presented itself in areas that were considered rural such as states in the south and in the west. Shipments on a large scale from the south and California were both made around the same time although natural ice was used from the Sierras in California rather than manufactured ice in the south. Refrigeration allowed for many areas to specialize in the growing of specific fruits. California specialized in several fruits, grapes, peaches, pears, plums, and apples while Georgia became famous for specifically its peaches. In California, the acceptance of the refrigerated rail carts lead to an increase of car loads from 4,500 carloads in 1895 to between 8,000 and 10,000 carloads in 1905. The Gulf States, Arkansas, Missouri and Tennessee entered into strawberry production on a large-scale while Mississippi became the center of the tomato industry. New Mexico, Colorado, Arizona, and Nevada grew cantaloupes. Without refrigeration this would have not been possible. By 1917, well-established fruit and vegetable areas that were close to eastern markets felt the pressure of competition from these distant specialized centers. Refrigeration was not limited to meat, fruit and vegetables but it also encompassed dairy product and dairy farms. In the early twentieth century large cities got their dairy supply from farms as far as 400 miles. Dairy products were not as easily transported great distances like fruits and vegetables due to greater perishability. Refrigeration made production possible in the west far from eastern markets, so much in fact that dairy farmers could pay transportation cost and still undersell their eastern competitors. Refrigeration and the refrigerated rail gave opportunity to areas with rich soil far from natural channel of transport such as a river, valley trail or harbors.

Rise of the Galactic City

"Edge city" was a term coined by Joel Garreau, whereas the term "galactic city" was coined by Lewis Mumford. These terms refer to a concentration of business, shopping, and entertainment outside a traditional downtown or central business district in what had previously been a residential or rural area. There were several factors contributing to the growth of these cities such as Los Angeles, Las Vegas, Houston, and Phoenix. The factors that contributed to these large cities include reliable automobiles, highway systems, refrigeration, and agricultural production increases. Large cities such as the ones mentioned above have not been uncommon in history but what separates these cities from the rest are that these cities are not along some natural channel of transport, or at some crossroad of two or more channels such as a trail, harbor, mountain, river, or valley. These large cities have been developed in areas that only a few hundred years ago would have been uninhabitable. Without a cost efficient way of cooling air and transporting water and food great distances these large cities would have never developed. The rapid growth of these cities was influenced by refrigeration and an agricultural productivity increase, allowing more distant farms to effectively feed the population.

Impact on Agriculture and Food Production

Agriculture's role in developed countries has drastically changed in the last century due to many factors, including refrigeration. Statistics from the 2007 census gives information on the large concentration of agricultural sales coming from a small portion of the existing farms in the United States today. This is a partial result of the market created for the frozen meat trade by the first successful shipment of frozen sheep carcasses coming from New Zealand in the 1880s. As the market continued to grow, regulations on food processing and quality began to be enforced. Eventually, electricity was introduced into rural homes in the United States, which allowed refrigeration technology to continue to expand on the farm, increasing output per person. Today, refrigeration's use on the farm reduces humidity levels, avoids spoiling due to bacterial growth, and assists in preservation.

Demographics

The introduction of refrigeration and evolution of additional technologies drastically changed agriculture in the United States. During the beginning of the 20th century, farming was a common occupation and lifestyle for United States citizens, as most farmers actually lived on their farm. In 1935, there were 6.8 million farms in the United States and a population of 127 million. Yet, while the United States population has continued to climb, citizens pursuing agriculture continue to decline. Based on the 2007 US Census, less than one percent of a population of 310 million people claim farming as an occupation today. However, the increasing population has led to an increasing demand for agricultural products, which is met through a greater variety of crops, fertilizers, pesticides, and improved technology. Improved technology has decreased the risk and time involved if agricultural management and allows larger farms to increase their output per person to meet society's demand.

Meat Packing and Trade

Prior to 1882, the South Island of New Zealand had been experimenting with sowing grass and crossbreeding sheep, which immediately gave their farmers economic potential in the exportation of meat. In 1882, the first successful shipment of sheep carcasses was sent from Port Chalmers in

Dunedin, New Zealand, to London. By the 1890s, the frozen meat trade became increasingly more profitable in New Zealand, especially in Canterbury, where 50% of exported sheep carcasses came from in 1900. It wasn't long before Canterbury meat was known for the highest quality, creating a demand for New Zealand meat around the world. In order to meet this new demand, the farmers improved their feed so sheep could be ready for the slaughter in only seven months. This new method of shipping led to an economic boom in New Zealand by the mid 1890s.

In the United States, the Meat Inspection Act of 1891 was put in place in the United States because local butchers felt the refrigerated railcar system was unwholesome. When meat packing began to take off, consumers became nervous about the quality of the meat for consumption. Upton Sinclair's 1906 novel *The Jungle* brought negative attention to the meat packing industry, by drawing to light unsanitary working conditions and processing of diseased animals. The book caught the attention of President Theodore Roosevelt, and the 1906 Meat Inspection Act was put into place as an amendment to the Meat Inspection Act of 1891. This new act focused on the quality of the meat and environment it is processed in.

Electricity in Rural Areas

In the early 1930s, 90 percent of the urban population of the United States had electric power, in comparison to only 10 percent of rural homes. At the time, power companies did not feel that extending power to rural areas (rural electrification) would produce enough profit to make it worth their while. However, in the midst of the Great Depression, President Franklin D. Roosevelt realized that rural areas would continue to lag behind urban areas in both poverty and production if they were not electrically wired. On May 11, 1935, the president signed an executive order called the Rural Electrification Administration, also known as REA. The agency provided loans to fund electric infrastructure in the rural areas. In just a few years, 300,000 people in rural areas of the United States had received power in their homes.

While electricity dramatically improved working conditions on farms, it also had a large impact on the safety of food production. Refrigeration systems were introduced to the farming and food distribution processes, which helped in food preservation and kept food supplies safe. Refrigeration also allowed for production of perishable commodities, which could then be shipped throughout the United States. As a result, the United States farmers quickly became the most productive in the world, and entire new food systems arose.

Farm Use

In order to reduce humidity levels and spoiling due to bacterial growth, refrigeration is used for meat, produce, and dairy processing in farming today. Refrigeration systems are used the heaviest in the warmer months for farming produce, which must be cooled as soon as possible in order to meet quality standards and increase the shelf life. Meanwhile, dairy farms refrigerate milk year round to avoid spoiling.

Effects on Lifestyle and Diet

In the late 19th Century and into the very early 20th Century, except for staple foods (sugar, rice, and beans) that needed no refrigeration, the available foods were affected heavily by the seasons and what could be grown locally. Refrigeration has removed these limitations. Refrigeration played a large part in the feasibility and then popularity of the modern supermarket. Fruits and vegetables out of season, or grown in distant locations, are now available at relatively low prices. Refrigerators have led to a huge increase in meat and dairy products as a portion of overall supermarket sales. As well as changing the goods purchased at the market, the ability to store these foods for extended periods of time has led to an increase in leisure time. Prior to the advent of the household refrigerator, people would have to shop on a daily basis for the supplies needed for their meals.

Impact on Nutrition

The introduction of refrigeration allowed for the hygienic handling and storage of perishables, and as such, promoted output growth, consumption, and the availability of nutrition. The change in our method of food preservation moved us away from salts to a more manageable sodium level. The ability to move and store perishables such as meat and dairy led to a 1.7% increase in dairy consumption and overall protein intake by 1.25% annually in the US after the 1890s.

People were not only consuming these perishables because it became easier for they themselves to store them, but because the innovations in refrigerated transportation and storage led to less spoilage and waste, thereby driving the prices of these products down. Refrigeration accounts for at least 5.1% of the increase in adult stature (in the US) through improved nutrition, and when the indirect effects associated with improvements in the quality of nutrients and the reduction in illness is additionally factored in, the overall impact becomes considerably larger. Recent studies have also shown a negative relationship between the number of refrigerators in a household and the rate of gastric cancer mortality.

Current Applications of Refrigeration

Probably the most widely used current applications of refrigeration are for air conditioning of private homes and public buildings, and refrigerating foodstuffs in homes, restaurants and large storage warehouses. The use of refrigerators in kitchens for storing fruits and vegetables has allowed adding fresh salads to the modern diet year round, and storing fish and meats safely for long periods. The optimum temperature range for perishable food storage is 3 to 5 °C (37 to 41 °F).

In commerce and manufacturing, there are many uses for refrigeration. Refrigeration is used to liquefy gases – oxygen, nitrogen, propane, and methane, for example. In compressed air purification, it is used to condense water vapor from compressed air to reduce its moisture content. In oil refineries, chemical plants, and petrochemical plants, refrigeration is used to maintain certain processes at their needed low temperatures (for example, in alkylation of butenes and butane to produce a high-octane gasoline component). Metal workers use refrigeration to temper steel and cutlery. When transporting temperature-sensitive foodstuffs and other materials by trucks, trains, airplanes and seagoing vessels, refrigeration is a necessity.

Dairy products are constantly in need of refrigeration, and it was only discovered in the past few decades that eggs needed to be refrigerated during shipment rather than waiting to be refrigerated after arrival at the grocery store. Meats, poultry and fish all must be kept in climate-controlled environments before being sold. Refrigeration also helps keep fruits and vegetables edible longer.

One of the most influential uses of refrigeration was in the development of the sushi/sashimi industry in Japan. Before the discovery of refrigeration, many sushi connoisseurs were at risk of contracting diseases. The dangers of unrefrigerated sashimi were not brought to light for decades due to the lack of research and healthcare distribution across rural Japan. Around mid-century, the Zojirushi corporation, based in Kyoto, made breakthroughs in refrigerator designs, making refrigerators cheaper and more accessible for restaurant proprietors and the general public.

Methods of Refrigeration

Methods of refrigeration can be classified as non-cyclic, cyclic, thermoelectric and magnetic.

Non-cyclic Refrigeration

This refrigeration method cools a contained area by melting ice, or by sublimating dry ice. Perhaps the simplest example of this is a portable cooler, where items are put in it, then ice is poured over the top. Regular ice can maintain temperatures near, but not below the freezing point, unless salt is used to cool the ice down further (as in a traditional ice-cream maker). Dry ice can reliably bring the temperature well below water freezing point.

Cyclic Refrigeration

This consists of a refrigeration cycle, where heat is removed from a low-temperature space or source and rejected to a high-temperature sink with the help of external work, and its inverse, the thermodynamic power cycle. In the power cycle, heat is supplied from a high-temperature source to the engine, part of the heat being used to produce work and the rest being rejected to a low-temperature sink. This satisfies the second law of thermodynamics.

A refrigeration cycle describes the changes that take place in the refrigerant as it alternately absorbs and rejects heat as it circulates through a refrigerator. It is also applied to heating, ventilation, and air conditioning HVACR work, when describing the "process" of refrigerant flow through an HVACR unit, whether it is a packaged or split system.

Heat naturally flows from hot to cold. Work is applied to cool a living space or storage volume by pumping heat from a lower temperature heat source into a higher temperature heat sink. Insulation is used to reduce the work and energy needed to achieve and maintain a lower temperature in the cooled space. The operating principle of the refrigeration cycle was described mathematically by Sadi Carnot in 1824 as a heat engine.

The most common types of refrigeration systems use the reverse-Rankine vapor-compression refrigeration cycle, although absorption heat pumps are used in a minority of applications.

Cyclic refrigeration can be classified as:

- Vapor cycle.
- Gas cycle.

Vapor cycle refrigeration can further be classified as:

- Vapor-compression refrigeration.

- Sorption Refrigeration:
    - Vapor-absorption refrigeration.
    - Adsorption refrigeration.

Vapor-compression Cycle

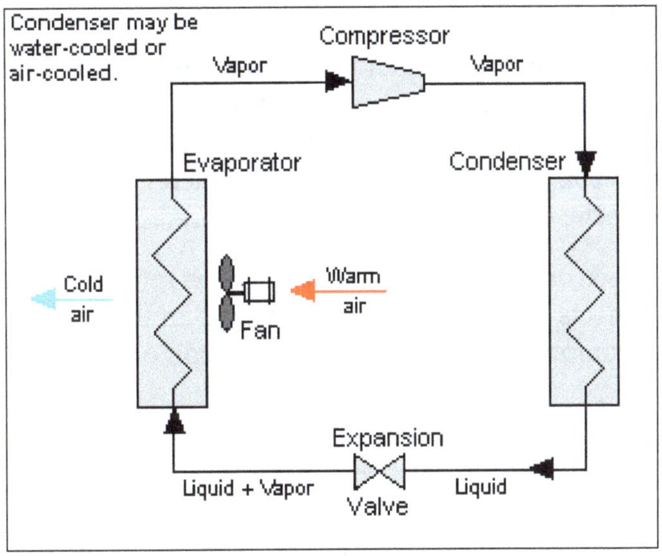

Vapor compression refrigeration.

The vapor-compression cycle is used in most household refrigerators as well as in many large commercial and industrial refrigeration systems. Figure provides a schematic diagram of the components of a typical vapor-compression refrigeration system.

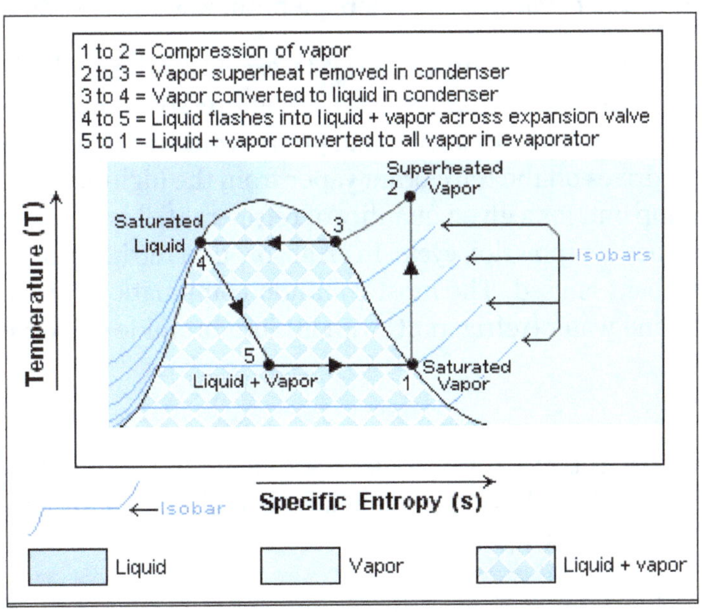

Temperature–Entropy diagram.

The thermodynamics of the cycle can be analyzed on a diagram as shown in figure. In this cycle, a circulating refrigerant such as Freon enters the compressor as a vapor. From point 1 to point 2, the vapor is compressed at constant entropy and exits the compressor as a vapor at a higher temperature, but still below the vapor pressure at that temperature. From point 2 to point 3 and on to point 4, the vapor travels through the condenser which cools the vapor until it starts condensing, and then condenses the vapor into a liquid by removing additional heat at constant pressure and temperature. Between points 4 and 5, the liquid refrigerant goes through the expansion valve (also called a throttle valve) where its pressure abruptly decreases, causing flash evaporation and auto-refrigeration of, typically, less than half of the liquid.

That results in a mixture of liquid and vapor at a lower temperature and pressure as shown at point 5. The cold liquid-vapor mixture then travels through the evaporator coil or tubes and is completely vaporized by cooling the warm air (from the space being refrigerated) being blown by a fan across the evaporator coil or tubes. The resulting refrigerant vapor returns to the compressor inlet at point 1 to complete the thermodynamic cycle.

The above discussion is based on the ideal vapor-compression refrigeration cycle, and does not take into account real-world effects like frictional pressure drop in the system, slight thermodynamic irreversibility during the compression of the refrigerant vapor, or non-ideal gas behavior, if any.

Sorption Cycle

Absorption Cycle

In the early years of the twentieth century, the vapor absorption cycle using water-ammonia systems or LiBr-water was popular and widely used. After the development of the vapor compression cycle, the vapor absorption cycle lost much of its importance because of its low coefficient of performance (about one fifth of that of the vapor compression cycle). Today, the vapor absorption cycle is used mainly where fuel for heating is available but electricity is not, such as in recreational vehicles that carry LP gas. It is also used in industrial environments where plentiful waste heat overcomes its inefficiency.

The absorption cycle is similar to the compression cycle, except for the method of raising the pressure of the refrigerant vapor. In the absorption system, the compressor is replaced by an absorber which dissolves the refrigerant in a suitable liquid, a liquid pump which raises the pressure and a generator which, on heat addition, drives off the refrigerant vapor from the high-pressure liquid. Some work is needed by the liquid pump but, for a given quantity of refrigerant, it is much smaller than needed by the compressor in the vapor compression cycle. In an absorption refrigerator, a suitable combination of refrigerant and absorbent is used. The most common combinations are ammonia (refrigerant) with water (absorbent), and water (refrigerant) with lithium bromide (absorbent).

Adsorption Cycle

In adsorption refrigeration, the refrigerant (adsorbate) could be ammonia, water, methanol, etc, while the adsorbent is a solid, such as silicone gel, activated carbon, or zeolite, unlike in the absorption cycle where absorbent is liquid.

Gas Cycle

When the working fluid is a gas that is compressed and expanded but doesn't change phase, the refrigeration cycle is called a gas cycle. Air is most often this working fluid. As there is no condensation and evaporation intended in a gas cycle, components corresponding to the condenser and evaporator in a vapor compression cycle are the hot and cold gas-to-gas heat exchangers in gas cycles.

The gas cycle is less efficient than the vapor compression cycle because the gas cycle works on the reverse Brayton cycle instead of the reverse Rankine cycle. As such the working fluid does not receive and reject heat at constant temperature. In the gas cycle, the refrigeration effect is equal to the product of the specific heat of the gas and the rise in temperature of the gas in the low temperature side. Therefore, for the same cooling load, a gas refrigeration cycle needs a large mass flow rate and is bulky.

Because of their lower efficiency and larger bulk, *air cycle* coolers are not often used nowadays in terrestrial cooling devices. However, the air cycle machine is very common on gas turbine-powered jet aircraft as cooling and ventilation units, because compressed air is readily available from the engines' compressor sections. Such units also serve the purpose of pressurizing the aircraft.

Thermoelectric Refrigeration

Thermoelectric cooling uses the Peltier effect to create a heat flux between the junction of two types of material. This effect is commonly used in camping and portable coolers and for cooling electronic components and small instruments.

Magnetic Refrigeration

Magnetic refrigeration, or adiabatic demagnetization, is a cooling technology based on the magnetocaloric effect, an intrinsic property of magnetic solids. The refrigerant is often a paramagnetic salt, such as cerium magnesium nitrate. The active magnetic dipoles in this case are those of the electron shells of the paramagnetic atoms.

A strong magnetic field is applied to the refrigerant, forcing its various magnetic dipoles to align and putting these degrees of freedom of the refrigerant into a state of lowered entropy. A heat sink then absorbs the heat released by the refrigerant due to its loss of entropy. Thermal contact with the heat sink is then broken so that the system is insulated, and the magnetic field is switched off. This increases the heat capacity of the refrigerant, thus decreasing its temperature below the temperature of the heat sink.

Because few materials exhibit the needed properties at room temperature, applications have so far been limited to cryogenics and research.

Other Methods

Other methods of refrigeration include the air cycle machine used in aircraft; the vortex tube used for spot cooling, when compressed air is available; and thermoacoustic refrigeration using sound waves in a pressurized gas to drive heat transfer and heat exchange; steam jet cooling popular in the early 1930s for air conditioning large buildings; thermoelastic cooling using a smart metal alloy stretching and relaxing. Many Stirling cycle heat engines can be run backwards to act as a refrigerator, and therefore these engines have a niche use in cryogenics. In addition there are other types of cryocoolers such as Gifford-McMahon coolers, Joule-Thomson coolers, pulse-tube refrigerators and, for temperatures between 2 mK and 500 mK, dilution refrigerators.

Elastocaloric Refrigeration

Another potential solid-state refrigeration technique and a relatively new area of study comes from a special property of super elastic materials. These materials undergo a temperature change when experiencing an applied mechanical stress (called the elastocaloric effect). Since super elastic materials deform reversibly at high strains, the material experiences a flattened elastic region in its stress-strain curve caused by a resulting phase transformation from an austenitic to a martensitic crystal phase.

When a super elastic material experiences a stress in the austenitic phase, it undergoes an exothermic phase transformation to the martensitic phase, which causes the material to heat up. Removing the stress reverses the process, restores the material to its austenitic phase, and absorbs heat from the surroundings cooling down the material.

The most appealing part of this research is how potentially energy efficient and environmentally friendly this cooling technology is. The different materials used, commonly shape-memory alloys, provide a non-toxic source of emission free refrigeration. The most commonly studied materials studied are shape-memory alloys, like nitinol and Cu-Zn-Al. Nitinol is of the more promising alloys with output heat at about 66 J/cm^3 and a temperature change of about 16–20 K. Due to the difficulty in manufacturing some of the shape memory alloys, alternative materials like natural rubber have been studied. Even though rubber may not give off as much heat per volume (12 J/cm^3) as the shape memory alloys, it still generates a comparable temperature change of about 12 K and operates at a suitable temperature range, low stresses, and low cost.

The main challenge however comes from potential energy losses in the form of hysteresis, often associated with this process. Since most of these losses comes from incompatibilities between the two phases, proper alloy tuning is necessary to reduce losses and increase reversibility and efficiency. Balancing the transformation strain of the material with the energy losses enables a large elastocaloric effect to occur and potentially a new alternative for refrigeration.

Fridge Gate

The Fridge Gate method is a theoretical application of using a single logic gate to drive a refrigerator in the most energy efficient way possible without violating the laws of thermodynamics. It operates on the fact that there are two energy states in which a particle can exist: the ground state and the excited state. The excited state carries a little more energy than the ground state, small enough so that the transition occurs with high probability. There are three components or particle types associated with the fridge gate. The first is on the interior of the fridge, the second on the outside and the third is connected to a power supply which heats up every so often that it can reach the E state and replenish the source. In the cooling step on the inside of the fridge, the g state particle absorbs energy from ambient particles, cooling them, and itself jumping to the e state. In the second step, on the outside of the fridge where the particles are also at an e state, the particle falls to the g state, releasing energy and heating the outside particles. In the third and final step, the power supply moves a particle at the e state, and when it falls to the g state it induces an energy-neutral swap where the interior e particle is replaced by a new g particle, restarting the cycle.

Passive Systems

MIT researchers have devised a new way of providing cooling on a hot sunny day, using inexpensive materials and requiring no fossil fuel-generated power. The passive system, which could be used to supplement other cooling systems to preserve food and medications in hot, off-grid locations, is essentially a high-tech version of a parasol.

Capacity Ratings

The measured capacity of refrigeration is always dimensioned in units of power. Domestic and commercial refrigerators may be rated in kJ/s, or Btu/h of cooling. For commercial and industrial refrigeration systems, the kilowatt (kW) is the basic unit of refrigeration except in North America, where the ton of refrigeration (TR) is used. [Nominally the capacity to freeze one short ton of water per day, the TR is defined as 12,000 Btu/hr (3.517 kW)].

A refrigeration system's coefficient of performance (CoP) is very important in determining a system's overall efficiency. It is defined as refrigeration capacity in kW divided by the energy input in kW. While CoP is a very simple measure of performance, it is typically not used for industrial refrigeration in North America. Owners and manufacturers of these systems typically use performance factor (PF). A system's PF is defined as a system's energy input in horsepower divided by its refrigeration capacity in TR. Both CoP and PF can be applied to either the entire system or to system components. For example, an individual compressor can be rated by comparing the energy needed to run the compressor versus the expected refrigeration capacity based on inlet volume flow rate. It is important to note that both CoP and PF for a refrigeration system are only defined at specific operating conditions, including temperatures and thermal loads. Moving away from the specified operating conditions can dramatically change a system's performance.

Sugaring

Sugaring is a method of food preservation that requires the food to be dehydrated and then to be packed with either crystallized sugar or with the liquids containing high amount of sugar such as honey or molasses.

The main purpose of this food preservation method is to treat the food in order to stop the growth of bacteria that may diminish the nutritional value and quality of the food. Some of the most popular sugared foods that are found in almost every cuisine across the globe include different kinds of fruits as well as a variety of vegetables such as ginger.

Process of Sugaring

Sugaring mainly reduces the action of the internal food enzymes by inhibiting the bacterial growth and this result in the preservation of food from spoiling.

The process of this type of food preservation is mainly done with any table or raw sugar that is generally in crystallized form. The sugar can also be liquefied in form of syrup, honey or molasses. These high sugar density liquids are of equal importance as sugar in the process of food preservation. Any fruit or vegetable that has to be preserved is washed thoroughly and desiccated by dehydration. The food is then cooked in liquid sugar product or raw sugar until crystallized and the resultant food is preserved in dried form.

Some fruits are even glazed in sugar syrup, but sold after being extracted from the liquid. The sugary coating on the fruits and the internal high sugar content increases the shelf life of the food.

Alcoholic preservation combined with sugar is also extremely popular to preserve some luxury items such as brandy and other spirits containing fruits. This type of food preservation should not be confused with the fruit flavored spirits like cherry brandy or apple wine.

Sugaring Food Preservation of various Foods

- Fruits – Apples, pears, plum, cherry, apricots and peaches are some of the popular fruits that are commonly preserved by sugaring method. These fruits are either dried before preservation or glazed in sugar syrup. Fruits are also combined with alcohol and preserved.
- Vegetables – Ginger and carrot are the most common vegetables that are often sugared to prepare relishes or sweet pickles. These candied vegetables are popularly served as condiments.
- Angelica – It is a herb that is widely used as a flavoring agent. However, sugar preserved or candied strips of angelica are extremely popular as cake decorations.
- Citrus peel – The peels of citrus fruits like lemon and amla (Indian gooseberry) are often candied to form relishes. However, 'murabba' (Indian candied dish) includes whole amla.

Advantages and Disadvantages of Sugaring Food Preservation Method

Sugaring has few advantages over other preservation methods, as this process does not require large number of ingredients and often the sugar extract or glaze is used to sweeten various other foods as well. It is also an easy preservation method with less time involvement. There is a risk in this method as sugar is believed to attract moisture very fast. When the atmospheric moisture is high in content, the yeast present in the environment starts its action and sugar starts fermenting into carbon-di-oxide and alcohol. Although fermented food is also a preserved food, the sugared foods should be prevented from fermenting, as it may lead to an unpleasant taste.

Frozen Food

A frozen processed foods aisle at a supermarket in Canada.

Freezing food preserves it from the time it is prepared to the time it is eaten. Since early times, farmers, fishermen, and trappers have preserved grains and produce in unheated buildings during the winter season. Freezing food slows down decomposition by turning residual moisture into ice, inhibiting the growth of most bacterial species. In the food commodity industry, there are two processes: mechanical and cryogenic (or flash freezing). The freezing kinetics is important to preserve the food quality and texture. Quicker freezing generates smaller ice crystals and maintains cellular structure. Cryogenic freezing is the quickest freezing technology available due to the ultra low liquid nitrogen temperature −196 °C (−320 °F).

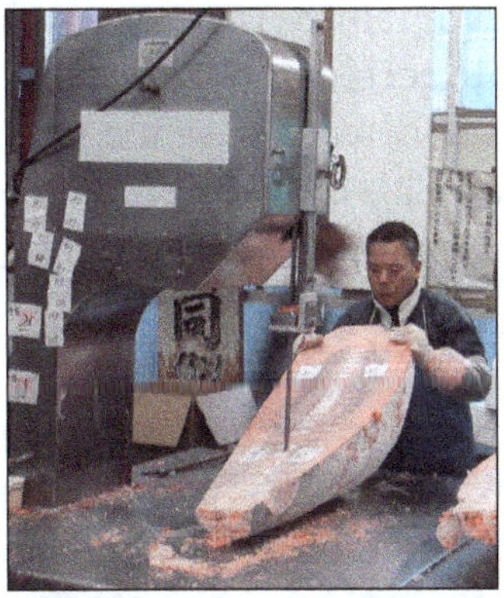

Cutting frozen tuna using a bandsaw in the Tsukiji fish market in Tokyo, Japan.

Preserving food in domestic kitchens during modern times is achieved using household freezers. Accepted advice to householders was to freeze food on the day of purchase. An initiative by a supermarket group in 2012 (backed by the UK's Waste & Resources Action Programme) promotes the freezing of food "as soon as possible up to the product's 'use by' date". The Food Standards Agency was reported as supporting the change, providing the food had been stored correctly up to that time.

Preservatives

Frozen products do not require any added preservatives because microorganisms do not grow when the temperature of the food is below −9.5 °C (15 °F), which is sufficient on its own in preventing food spoilage. Long-term preservation of food may call for food storage at even lower temperatures. Carboxymethylcellulose (CMC), a tasteless and odorless stabilizer, is typically added to frozen food because it does not adulterate the quality of the product.

Technology

The freezing technique itself, just like the frozen food market, is developing to become faster, more efficient and more cost-effective.

Mechanical freezers were the first to be used in the food industry and are used in the vast majority of freezing/refrigerating lines. They function by circulating a refrigerant, normally ammonia, around the system, which withdraws heat from the food product. This heat is then transferred to a condenser and dissipated into air or water. The refrigerant itself, now a high pressure, hot liquid, is directed into an evaporator. As it passes through an expansion valve, it is cooled and then vaporises into a gaseous state. Now a low pressure, low temperature gas again, it can be reintroduced into the system.

Cryogenic or (flash freezing) of food is a more recent development, but is used by many leading food manufacturers all over the world. Cryogenic equipment uses very low temperature gases – usually liquid nitrogen or solid carbon dioxide – which are applied directly to the food product.

Packaging

Frozen food packaging must maintain its integrity throughout filling, sealing, freezing, storage, transportation, thawing, and often cooking. As many frozen foods are cooked in a microwave oven, manufacturers have developed packaging that can go straight from freezer to the microwave.

In 1974, the first differential heating container (DHC) was sold to the public. A DHC is a sleeve of metal designed to allow frozen foods to receive the correct amount of heat. Various sized apertures were positioned around the sleeve. The consumer would put the frozen dinner into the sleeve according to what needed the most heat. This ensured proper cooking.

Today there are multiple options for packaging frozen foods. Boxes, cartons, bags, pouches, Boil-in-Bags, lidded trays and pans, crystallized PET trays, and composite and plastic cans.

Scientists are continually researching new aspects of frozen food packaging. Active packaging offers a host of new technologies that can actively sense and then neutralize the presence of bacteria or other harmful species. Active packaging can extend shelf-life, maintain product safety, and help preserve the food over a longer period of time. Several functions of active packaging are being researched:

- Oxygen scavengers,
- Time Temperature Indicators and digital temperature data loggers,
- Antimicrobials,
- Carbon Dioxide controllers,
- Microwave susceptors,
- Moisture control: Water activity, Moisture vapor transmission rate, etc.,
- Flavor enhancers,
- Odor generators,
- Oxygen-permeable films,
- Oxygen generators.

Effects on Nutrients

Vitamin Content of Frozen Foods

- Vitamin C: Usually lost in a higher concentration than any other vitamin. A study was performed on peas to determine the cause of vitamin C loss. A vitamin loss of ten percent occurred during the blanching phase with the rest of the loss occurring during the cooling and washing stages. The vitamin loss was not actually accredited to the freezing process. Another experiment was performed involving peas and lima beans. Frozen and canned vegetables were both used in the experiment. The frozen vegetables were stored at −23 °C (−10 °F) and the canned vegetables were stored at room temperature 24 °C (75 °F). After 0, 3, 6, and 12 months of storage, the vegetables were analyzed with and without cooking. O'Hara, the scientist performing the experiment said, "From the view point of the vitamin content of the two vegetables when they were ready for the plate of the consumer, there did not appear to be any marked advantages attributable to method of preservation, frozen storage, processed in a tin, or processed in glass".

- Vitamin B_1 (Thiamin): A vitamin loss of 25 percent is normal. Thiamin is easily soluble in water and is destroyed by heat.
- Vitamin B_2 (Riboflavin): Not much research has been done to see how much freezing affects Riboflavin levels. Studies that have been performed are inconclusive; one study found an 18 percent vitamin loss in green vegetables, while another determined a 4 percent loss. It is commonly accepted that the loss of Riboflavin has to do with the preparation for freezing rather than the actual freezing process itself.
- Vitamin A (Carotene): There is little loss of carotene during preparation for freezing and freezing of most vegetables. Much of the vitamin loss is incurred during the extended storage period.

Effectiveness

A frozen food warehouse at McMurdo Station.

Freezing is an effective form of food preservation because the pathogens that cause food spoilage are killed or do not grow very rapidly at reduced temperatures. The process is less effective in food preservation than are thermal techniques, such as boiling, because pathogens are more likely to be able to survive cold temperatures rather than hot temperatures. One of the problems surrounding the use of freezing as a method of food preservation is the danger that pathogens deactivated (but not killed) by the process will once again become active when the frozen food thaws.

Foods may be preserved for several months by freezing. Long-term frozen storage requires a constant temperature of −18 °C (0 °F) or less.

Defrosting

To be used, many cooked foods that have been previously frozen require defrosting prior to consumption. Preferably, some frozen meats should be defrosted prior to cooking to achieve the best outcome: cooked through evenly and of good texture.

Ideally, most frozen foods should be defrosted in a refrigerator to avoid significant growth of pathogens. However, this can take considerable time.

Food is often defrosted in one of several ways:

- At room temperature; this is dangerous since the outside may be defrosted while the inside remains frozen.
- In a refrigerator.
- In a microwave oven.
- Wrapped in plastic and placed in cold water or under cold running water.

People sometimes defrost frozen foods at room temperature because of time constraints or ignorance; such foods should be promptly consumed after cooking or discarded and never be refrozen or refrigerated since pathogens are not killed by the freezing process.

Quality

The speed of the freezing has a direct impact on the size and the number of ice crystals formed within a food product's cells and extracellular space. Slow freezing leads to fewer but larger ice crystals while fast freezing leads to smaller but more numerous ice crystals. Large ice crystals can puncture the walls of the cells of the food product which will cause a degradation of the texture of the product as well as the loss of its natural juices during thawing. That is why there will be a qualitative difference observed between food products frozen by ventilated mechanical freezing, non-ventilated mechanical freezing or cryogenic freezing with liquid nitrogen.

Reaction: According to a study, an American consumes on average 71 frozen foods a year, most of which are pre-cooked frozen meals.

Canning

Special-edition steel soup cans commemorating Andy Warhol's paintings.

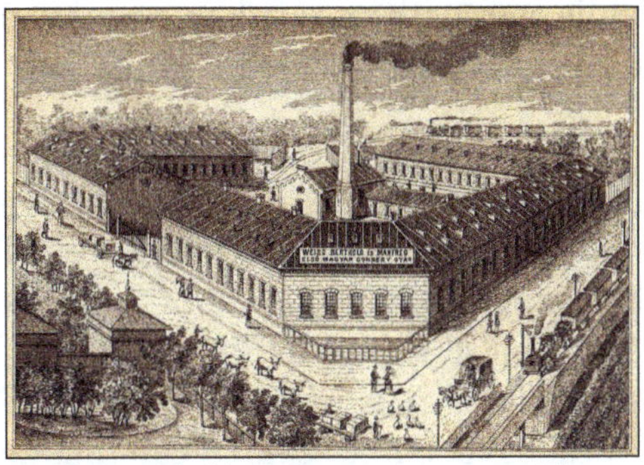

The Berthold-Weiss Factory, one of the first large canned food factories in Csepel-Budapest.

Canning is a method of preserving food in which the food contents are processed and sealed in an airtight container (jars like Mason jars, and steel and tin cans). Canning provides a shelf life typically ranging from one to five years, although under specific circumstances it can be much longer. A freeze-dried canned product, such as canned dried lentils, could last as long as 30 years in an edible state. In 1974, samples of canned food from the wreck of the *Bertrand*, a steamboat that sank in the Missouri River in 1865, were tested by the National Food Processors Association. Although appearance, smell and vitamin content had deteriorated, there was no trace of microbial growth and the 109-year-old food was determined to be still safe to eat.

Methods

The original fragile and heavy glass containers presented challenges for transportation, and glass jars were largely replaced in commercial canneries with cylindrical tin can or wrought-iron canisters (later shortened to "cans") following the work of Peter Durand. Cans are cheaper and quicker to make, and much less fragile than glass jars. Glass jars have remained popular for some high-value products and in home canning. Can openers were not invented for another thirty years – at first, soldiers had to cut the cans open with bayonets or smash them open with rocks. Today, tin-coated steel is the material most commonly used. Laminate vacuum pouches are also used for canning, such as used in MREs and Capri Sun drinks.

To prevent the food from being spoiled before and during containment, a number of methods are used: Pasteurisation, boiling (and other applications of high temperature over a period of time), refrigeration, freezing, drying, vacuum treatment, antimicrobial agents that are natural to the recipe of the foods being preserved, a sufficient dose of ionizing radiation, submersion in a strong saline solution, acid, base, osmotically extreme (for example very sugary) or other microbially-challenging environments.

Other than sterilization, no method is perfectly dependable as a preservative. For example, the microorganism *Clostridium botulinum* (which causes botulism) can be eliminated only at temperatures above the boiling point of water.

From a public safety point of view, foods with low acidity (a pH more than 4.6) need sterilization under high temperature (116–130 °C). To achieve temperatures above the boiling point requires the use of a pressure canner. Foods that must be pressure canned include most vegetables, meat, seafood, poultry, and dairy products. The only foods that may be safely canned in an ordinary boiling water bath are highly acidic ones with a pH below 4.6, such as fruits, pickled vegetables, or other foods to which acidic additives have been added.

Double Seams

Invented in 1888 by Max Ams, modern double seams provide an airtight seal to the tin can. This airtight nature is crucial to keeping micro-organisms out of the can and keeping its contents sealed inside. Thus, double seamed cans are also known as Sanitary Cans. Developed in 1900 in Europe, this sort of can was made of the traditional cylindrical body made with tin plate. The two ends (lids) were attached using what is now called a double seam. A can thus sealed is impervious to contamination by creating two tight continuous folds between the can's cylindrical body and the lids. This eliminated the need for solder and allowed improvements in manufacturing speed, reducing cost.

Double seaming uses rollers to shape the can, lid and the final double seam. To make a sanitary can and lid suitable for double seaming, manufacture begins with a sheet of coated tin plate. To create the can body, rectangles are cut and curled around a die, and welded together creating a cylinder with a side seam.

Rollers are then used to flare out one or both ends of the cylinder to create a quarter circle flange around the circumference. Precision is required to ensure that the welded sides are perfectly aligned, as any misalignment will cause inconsistent flange shape, compromising its integrity.

A circle is then cut from the sheet using a die cutter. The circle is shaped in a stamping press to create a downward countersink to fit snugly into the can body. The result can be compared to an upside down and very flat top hat. The outer edge is then curled down and around about 140 degrees using rollers to create the end curl.

The result is a steel tube with a flanged edge, and a countersunk steel disc with a curled edge. A rubber compound is put inside the curl.

Seaming

Opened can.

The body and end are brought together in a seamer and held in place by the base plate and chuck, respectively. The base plate provides a sure footing for the can body during the seaming operation and the chuck fits snugly into the end (lid). The result is the countersink of the end sits inside the top of the can body just below the flange. The end curl protrudes slightly beyond the flange.

First Operation

Once brought together in the seamer, the seaming head presses a first operation roller against the end curl. The end curl is pressed against the flange curling it in toward the body and under the flange. The flange is also bent downward, and the end and body are now loosely joined together. The first operation roller is then retracted. At this point five thicknesses of steel exist in the seam. From the outside in they are:

- End
- Flange
- End Curl
- Body
- Countersink

Second Operation

The seaming head then engages the second operation roller against the partly formed seam. The second operation presses all five steel components together tightly to form the final seal. The five layers in the final seam are then called; a) End, b) Body Hook, c) Cover Hook, d) Body, e) Countersink. All sanitary cans require a filling medium within the seam because otherwise the metal-to-metal contact will not maintain a hermetic seal. In most cases, a rubberized compound is placed inside the end curl radius, forming the critical seal between the end and the body.

Probably the most important innovation since the introduction of double seams is the welded side seam. Prior to the welded side seam, the can body was folded and/or soldered together, leaving a relatively thick side seam. The thick side seam required that the side seam end juncture at the end curl to have more metal to curl around before closing in behind the Body Hook or flange, with a greater opportunity for error.

Seamer Setup and Quality Assurance

Many different parts during the seaming process are critical in ensuring that a can is airtight and vacuum sealed. The dangers of a can that is not hermetically sealed are contamination by foreign objects (bacteria or fungicide sprays), or that the can could leak or spoil.

One important part is the seamer setup. This process is usually performed by an experienced technician. Amongst the parts that need setup are seamer rolls and chucks which have to be set in their exact position (using a feeler gauge or a clearance gauge). The lifter pressure and position, roll and chuck designs, tooling wear, and bearing wear all contribute to a good double seam.

Incorrect setups can be non-intuitive. For example, due to the springback effect, a seam can appear loose, when in reality it was closed too tight and has opened up like a spring. For this reason, experienced operators and good seamer setup are critical to ensure that double seams are properly closed.

Quality control usually involves taking full cans from the line – one per seamer head, at least once or twice per shift, and performing a teardown operation (wrinkle/tightness), mechanical tests (external thickness, seamer length/height and countersink) as well as cutting the seam open with a twin blade saw and measuring with a double seam inspection system. The combination of these measurements will determine the seam's quality.

Use of a statistical process control (SPC) software in conjunction with a manual double-seam monitor, computerized double seam scanner, or even a fully automatic double seam inspection system makes the laborious process of double seam inspection faster and much more accurate. Statistically tracking the performance of each head or seaming station of the can seamer allows for better prediction of can seamer issues, and may be used to plan maintenance when convenient, rather than to simply react after bad or unsafe cans have been produced.

Nutritional Value

Canning is a way of processing food to extend its shelf life. The idea is to make food available and edible long after the processing time. A 1997 study found that canned fruits and vegetables are as rich with dietary fiber and vitamins as the same corresponding fresh or frozen foods, and in some cases the canned products are richer than their fresh or frozen counterparts. The heating process during canning appears to make dietary fiber more soluble, and therefore more readily fermented in the colon into gases and physiologically active byproducts. Canned tomatoes have a higher available lycopene content. Consequently, canned meat and vegetables are often among the list of food items that are stocked during emergencies.

Potential Hazards

Women working in a cannery.

In the beginning of the 19th century the process of canning foods was mainly done by small canneries. These canneries were full of overlooked sanitation problems, such as poor hygiene and unsanitary work environments. Since the refrigerator did not exist and industrial canning standards were not set in place it was very common for contaminated cans to slip onto the grocery store shelves.

Migration of can Components

In canning toxicology, migration is the movement of substances from the can itself into the contents. Potential toxic substances that can migrate are lead, causing lead poisoning, or bisphenol A (BPA), a potential endocrine disruptor that is an ingredient in the epoxy commonly used to coat the inner surface of cans. Some cans are manufactured with a BPA-free enamel lining produced from plant oils and resins. On 20 February 2018, Packaging Digest reported that "At least 90%" of food cans no longer contained BPA.

Salt Content

Salt (sodium chloride), dissolved in water, is used in the canning process. As a result, canned food can be a major source of dietary salt. Too much salt increases the risk of health problems, including high blood pressure. Therefore, health authorities have recommended limitations of dietary sodium. Many canned products are available in low-salt and no-salt alternatives.

Rinsing thoroughly after opening may reduce the amount of salt in canned foods, since much of the salt content is thought to be in the liquid, rather than the food itself.

Botulism

Foodborne botulism results from contaminated foodstuffs in which *C. botulinum* spores have been allowed to germinate and produce botulism toxin, and this typically occurs in canned non-acidic food substances that have not received a strong enough thermal heat treatment. *C. botulinum* prefers low oxygen environments and is a poor competitor to other bacteria, but its spores are resistant to thermal treatments. When a canned food is sterilized insufficiently, most other bacteria besides the *C. botulinum* spores are killed, and the spores can germinate and produce botulism toxin. Botulism is a rare but serious paralytic illness, leading to paralysis that typically starts with the muscles of the face and then spreads towards the limbs. The botulinum toxin is extremely dangerous because it cannot be detected by sight or smell, and ingestion of even a small amount of the toxin can be deadly. In severe forms, it leads to paralysis of the breathing muscles and causes respiratory failure. In view of this life-threatening complication, all suspected cases of botulism are treated as medical emergencies, and public health officials are usually involved to prevent further cases from the same source.

Canning and Economic Recession

Canned goods and canning supplies sell particularly well in times of recession due to the tendency of financially stressed individuals to engage in cocooning, a term used by retail analysts to describe the phenomenon in which people choose to stay at home instead of adding expenditures to their budget by dining out and socializing outside the home.

In February 2009 during a recession, the United States saw an 11.5% rise in sales of canning-related items.

Some communities in the US have county canning centers which are available for teaching canning, or shared community kitchens which can be rented for canning one's own foods.

Food Drying

Flattened fish drying in the sun in Madagascar. Fish are preserved through such traditional methods as drying, smoking and salting.

A whole potato, sliced pieces (right), and dried sliced pieces (left).

Food drying is a method of food preservation in which food is dried (dehydrated or desiccated). Drying inhibits the growth of bacteria, yeasts, and mold through the removal of water. Dehydration has been used widely for this purpose since ancient times; the earliest known practice is 12,000 B.C. by inhabitants of the modern Middle East and Asia regions. Water is traditionally removed through evaporation (air drying, sun drying, smoking or wind drying), although today electric food dehydrators or freeze-drying can be used to speed the drying process and ensure more consistent results.

Food Types

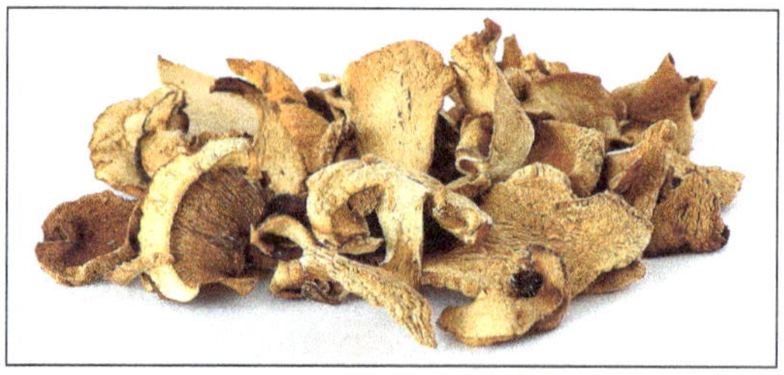

A collection of dried mushrooms.

Many different foods can be prepared by dehydration. Meat has held a historically significant role. For centuries, much of the European diet depended on dried cod—known as salt cod, bacalhau (with salt), or stockfish (without). It formed the main protein source for the slaves on the West Indian plantations, and was a major economic force within the triangular trade. Dried fish most commonly cod or haddock, known as Harðfiskur, is a delicacy in Iceland, while dried reindeer meat is a traditional Sami food. Dried meats include prosciutto (Parma ham), bresaola, biltong and beef jerky.

Sun-drying octopus.

Dried fruits have been consumed historically due to their high sugar content and sweet taste, and a longer shelf-life from drying. Fruits may be used differently when dried. The plum becomes a prune, the grape a raisin. Figs and dates may be transformed into different products that can either be eaten as they are, used in recipes, or rehydrated.

Freeze-dried vegetables are often found in food for backpackers, hunters, and the military. Garlic and onion are often dried. Edible mushrooms, as well as other fungi, are also sometimes dried for preservation purposes or to be used as seasonings.

Preparation

Home drying of vegetables, fruit and meat can be carried out with electrical dehydrators (household appliance) or by sun-drying or by wind. Preservatives such as potassium metabisulfite, BHA, or BHT may be used, but are not required. However, dried products without these preservatives may require refrigeration or freezing to ensure safe storage for a long time.

Industrial food dehydration is often accomplished by freeze-drying. In this case food is flash frozen and put into a reduced-pressure system which causes the water to sublimate directly from the solid to the gaseous phase. Although freeze-drying is more expensive than traditional dehydration techniques, it also mitigates the change in flavor, texture, and nutritional value.In addition, another widely used industrial method of drying of food is convective hot air drying. Industrial hot air dryers are simple and easy to design, construct and maintain. More so, it is very affordable and has been reported to retain most of the nutritional properties of food if dried using appropriate drying conditions.

Another form of food dehydration is irradiation. Irradiation uses x-rays, ultraviolet light, and ionizing radiations to penetrate food to the point of sterilization. Astronauts and people who are highly at risk for microbial infections benefit from this method of food drying.

Hurdle technology is the combination of multiple food preservation methods. Hurdle technology uses low doses of multiple food preservation techniques in order to ensure food is not only safe but is desirable visually and texturally.

Packaging

Packaging ensures effective food preservation. Some methods of packaging that are beneficial to dehydrated food are vacuum sealed, inert gases, or gases that help regulate respiration, biological organisms, and growth of microorganisms.

Other Methods

This electric food dehydrator has a hot air blower that blows air through trays with foods on them. Pictured are mango and papaya slices being dried.

There are many different methods for drying, each with their own advantages for particular applications. These include:

- Convection drying,
- Bed dryers,
- Drum drying,
- Freeze Drying,
- Microwave-vacuum drying,
- Shelf dryers,
- Spray drying,
- Infrared radiation drying,
- Combined thermal hybrid drying,
- Sunlight,
- Commercial food dehydrators,
- Household oven.

Fruit Preserves, Jams and Jellies

The making of jellies and other preserves is an old and popular process, providing a means of keeping fruits far beyond their normal storage life and sometimes making use of blemished or off-grade fruits that may not be ideal for fresh consumption. In jelly making, the goal is to produce a clear, brilliant gel from the juice of a chosen fruit. Jams are made from the entire fruit, including the pulp, while preserves are essentially jellies that contain whole or large pieces. Marmalade, usually made from citrus fruit, is a jellylike concentrate of prepared juice and sliced peel.

The essential ingredients for a successful preserve are sugar, acid, and pectin. These three ingredients lower the pH of the preserve and bind available water, thus creating an environment in which the growth of microorganisms is retarded. In some cases the fruit can provide all the pectin and acid that are needed. If the acid content of the fruit is low, external sources such as lemon juice can be added. Similarly, if the planned mix of fruit is low in pectin, a commercial source may be used. Sugar is always added, and in general all of the three essential ingredients have to be added in order to create a successful product.

The making of preserves begins with an initial mix containing not less than 45 parts by weight fruit for every 55 parts by weight sugar solids. The sugar solids are added after the fruit is crushed, and the mix is then cooked. Cooking may be done in a highly controlled vacuum kettle, in which flavour volatiles are captured and returned to the product. The cooking process continues until the heated mix is concentrated to a predetermined level of soluble solids. A generally accepted level is 65 percent soluble solids; at this concentration the boiling temperature is 7° to 12° above the boiling point of water. The product is then transferred to containers and sealed as a shelf-stable product.

The exact amount of sugar needed depends on the acidity level, the natural sugar content, and the type of product desired. If sugar content is too low, the resulting jelly will be tough; excessive sugar, on the other hand, will create a "soft set" that can be broken easily. Appropriate amounts of acid and pectin are added during the cooking process. The pH must be adjusted to an acidic level of approximately 3.1. Increased acidity reduces the amount of sugar needed in the blend, although excessive acidity can cause syneresis, or a separation of liquid from the gel. If the pectin level is inadequate, then the preserve will not "set"; that is, not enough water will be bound to create a complete gel.

Fruit Preservation

Since fruits are generally acidic, they are naturally amenable to preservation. The premier role of acidity in preservation is to stop bacterial growth. Second, increased acidity can activate chemical reactions such as pectin set, which lowers water activity and reduces the possibility of microbial growth.

Dehydration

Dehydration is among the oldest and most common forms of fruit preservation. In dehydration, moisture in the fruit is driven off, leaving a stable food that has a moisture content below that at which microorganisms can grow. There are three basic systems for dehydration: sun drying, such as that used for raisins; hot-air dehydration; and freeze-drying.

Dehydration has a number of advantages. Dehydrated fruit has a virtually unlimited shelf life when held under proper storage conditions. Drying does not significantly reduce the calories or minerals, and vitamin losses are similar to other preservation methods. In addition, by reducing the weight and the need for refrigeration, handling and transportation costs can be reduced dramatically. Dehydrated fruits are typically reduced in weight by 75 to 90 percent.

Although dehydration offers a convenient product form, it usually requires a careful inactivation of enzymes. This is usually accomplished by blanching of the fruit or by chemical inactivation. Typically, sulfur dioxide is added for its antioxidant and preservative effects. In order to control browning, the fruit is often treated prior to dehydration with sodium sulfite and sodium bisulfite.

Thermal Processes

In thermal processing, heat is used to destroy spoilage organisms and to inactivate troublesome enzymes. Enzymes are typically responsible for browning, softening, and the development of off-flavours. For high-acid fruit products the most typical thermal process is canning, in which fruit or fruit products are hot-filled or heated in a hermetically sealed container. The process temperature is generally in the range of 88 °C (190 °F).

Chemical Preservation

Chemicals also can be used as a preservative, either through artificial addition or through the action of microorganisms. An example of the latter method is yeast fermentation, which can cause an increase in ethyl alcohol sufficient to preserve the fruit product. Pickling is another example of chemical preservation. In the case of pickling, the product may be preserved by the addition of salt, sugar, acetic acid (vinegar), and alcohol. High sugar content also acts as a fruit preservative by tying up all available moisture so that microorganisms cannot grow.

Irradiation

Although irradiation is an expensive method, it has been shown to be an effective means of extending the shelf life of fresh fruits. Irradiated fruit products have not been well received by the public, even in light of evidence supporting the healthfulness and safety of such foods.

Freezing

Freezing of fruits and fruit products is a common consumer practice. Cold temperatures act to retard the spoilage of fruit by inhibiting microbial action and slowing metabolic processes. In order to achieve extended storage life, the product must be held well below the freezing point of water—typically at a minimum of -23 °C (-10 °F). Generally, rapid freezing leads to an improved texture upon thawing.

A prerequisite for effective freezing is inactivation of fruit enzymes. Traditionally, this is done through blanching or by the addition of a chemical. Blanching consists of heating the fruit for a short time in water or steam prior to cooling and subsequent freezing. The blanch step is intended to inactivate enzyme systems responsible for off-flavours, browning, and softening.

Smoking

Smoking in food processing is the exposure of cured meat and fish products to smoke for the purposes of preserving them and increasing their palatability by adding flavour and imparting a rich brown colour. The drying action of the smoke tends to preserve the meat, though many of the chemicals present in wood smoke (e.g., formaldehyde and certain alcohols) are natural preservatives as well.

Smoking is one of the oldest of food preservation methods, probably having arisen shortly after the development of cooking with fire. The practice attained high levels of sophistication in several cultures, notably the smoking of fish in Scandinavia and northwestern North America and the production of smoked hams in Europe and the United States. Interest in smoking meats, which had declined during the mid-20th century owing to the popularity of chemical preservatives, was revived late in the century by the so-called natural or health food movement.

Whether done on a commercial or a home scale, the smoking technique involves hanging the meat or placing it on racks in a chamber designed to contain the smoke. Commercial smokehouses, usually several stories high, often use steampipes to supplement the heat of a natural sawdust fire. Hickory sawdust is the preferred fuel. Whatever the size of the smoking operation, it is imperative that a hardwood fire be used. The softwood of conifers such as spruce and pine contains pitch, which produces a film on the meat and imparts a bitter taste. Generally, smokehouse temperatures vary from 109 to 160 °F (43 to 71 °C), and smoking periods vary from as short as a few hours to as long as several days, depending on the type of meat and its moisture content. After smoking, the meat is chilled as rapidly as possible and cut and wrapped for the retail trade.

In the United States, pork and beef hams, bacon bellies, and sausages are the most common commercially smoked meats. However, amateurs using ordinary smoke ovens or adapting barbecue grills to the purpose have successfully used the smoking technique to flavour and preserve not only meat, fowl, and fish but also cheeses, nuts and seeds, hard-boiled eggs, and berries, as well as the variety meats including heart, tongue, and liver.

In order to shorten the production process, commercial meats are sometimes artificially "smoked" by dipping them in a solution of preservative chemicals or by painting them with such a solution. But because this procedure involves no natural drying action, it has practically no preservative effect.

References

- Salt-preserve-food: preserveandpickle.com, Retrieved 26 July, 2019
- Bitterman, m. (2010). "salt reference guide". Salted: a manifesto on the world's most essential mineral, with recipes. Random house. P. 187. Isbn 978-1580082624. Retrieved 2013-08-16
- Parthasarathy dk1, bryan ns; bryan (2004). "sodium nitrite: the "cure" for nitric oxide insufficiency". Meat science. 92 (3): 274–279. Doi:10.1016/j.meatsci.2012.03.001. Pmid 22464105
- "Pickle bill fact sheet". 13 march 2008. Archived from the original on 13 march 2008. Retrieved 15 february 2018
- Fruit-preserves-jams-and-jellies, fruit-processing: britannica.com, Retrieved 21 May, 2019
- Mcgee, harold (2004). On food and cooking: the science and lore of the kitchen. New york: scribner, pp. 291–296. Isbn 0-684-80001-2

5
Food Pathogens

The parasites in food that cause illness or food poisoning are referred to as food pathogens. Some of the plant pathogens are Escherichia coli O157:H7, Campylobacter, Clostridium perfringens, Listeria monocytogenes, Norovirus, Toxoplasmosis, Giardiasis, etc. This chapter discusses these food pathogens in detail.

Escherichia Coli O157:H7

Escherichia coli (E. coli) is a bacterium that is commonly found in the gut of humans and warm-blooded animals. Most strains of E. coli are harmless. Some strains however, such as Shiga toxin-producing E. coli (STEC), can cause severe foodborne disease. It is transmitted to humans primarily through consumption of contaminated foods, such as raw or undercooked ground meat products, raw milk, and contaminated raw vegetables and sprouts.

STEC produces toxins, known as Shiga-toxins because of their similarity to the toxins produced by Shigella dysenteriae. STEC can grow in temperatures ranging from 7 °C to 50 °C, with an optimum temperature of 37 °C. Some STEC can grow in acidic foods, down to a pH of 4.4, and in foods with a minimum water activity (a_w) of 0.95.

STEC is destroyed by thorough cooking of foods until all parts reach a temperature of 70 °C or higher. E. coli O157:H7 is the most important STEC serotype in relation to public health; however, other serotypes have frequently been involved in sporadic cases and outbreaks.

Symptoms

Symptoms of the diseases caused by STEC include abdominal cramps and diarrhoea that may in some cases progress to bloody diarrhoea (haemorrhagic colitis). Fever and vomiting may also occur. The incubation period can range from 3 to 8 days, with a median of 3 to 4 days. Most patients recover within 10 days, but in a small proportion of patients (particularly young children and the elderly), the infection may lead to a life-threatening disease, such as haemolytic uraemic syndrome (HUS). HUS is characterized by acute renal failure, haemolytic anaemia and thrombocytopenia (low blood platelets).

It is estimated that up to 10% of patients with STEC infection may develop HUS, with a case-fatality rate ranging from 3 to 5%. Overall, HUS is the most common cause of acute renal failure in young children. It can cause neurological complications (such as seizure, stroke and coma) in 25% of HUS patients and chronic renal sequelae, usually mild, in around 50% of survivors.

Persons who experience bloody diarrhoea or severe abdominal cramps should seek medical care. Antibiotics are not part of the treatment of patients with STEC disease and may possibly increase the risk of subsequent HUS.

Sources and Transmission

Most available information on STEC relates to serotype O157:H7, since it is easily differentiated biochemically from other E. coli strains. The reservoir of this pathogen appears to be mainly cattle. In addition, other ruminants such as sheep, goats, deer are considered significant reservoirs, while other mammals (such as pigs, horses, rabbits, dogs, and cats) and birds (such as chickens and turkeys) have been found infected.

E. coli O157:H7 is transmitted to humans primarily through consumption of contaminated foods, such as raw or undercooked ground meat products and raw milk. Faecal contamination of water and other foods, as well as cross-contamination during food preparation (with beef and other meat products, contaminated surfaces and kitchen utensils), will also lead to infection. Examples of foods implicated in outbreaks of E. coli O157:H7 include undercooked hamburgers, dried cured salami, unpasteurized fresh-pressed apple cider, yogurt, and cheese made from raw milk.

An increasing number of outbreaks are associated with the consumption of fruits and vegetables (including sprouts, spinach, lettuce, coleslaw, and salad) whereby contamination may be due to contact with faeces from domestic or wild animals at some stage during cultivation or handling. STEC has also been isolated from bodies of water (such as ponds and streams), wells and water troughs, and has been found to survive for months in manure and water-trough sediments. Waterborne transmission has been reported, both from contaminated drinking-water and from recreational waters.

Person-to-person contact is an important mode of transmission through the oral-faecal route. An asymptomatic carrier state has been reported, where individuals show no clinical signs of disease but are capable of infecting others. The duration of excretion of STEC is about 1 week or less in adults, but can be longer in children. Visiting farms and other venues where the general public might come into direct contact with farm animals has also been identified as an important risk factor for STEC infection.

Prevention

The prevention of infection requires control measures at all stages of the food chain, from agricultural production on the farm to processing, manufacturing and preparation of foods in both commercial establishments and household kitchens.

Industry

The number of cases of disease might be reduced by various mitigation strategies for ground beef (for example, screening the animals pre-slaughter to reduce the introduction of large numbers of pathogens in the slaughtering environment). Good hygienic slaughtering practices reduce contamination of carcasses by faeces, but do not guarantee the absence of STEC from products.

Education in hygienic handling of foods for workers at farms, abattoirs and those involved in the food production is essential to keep microbiological contamination to a minimum. The only effective method of eliminating STEC from foods is to introduce a bactericidal treatment, such as heating (for example, cooking or pasteurization) or irradiation.

Household

Preventive measures for E. coli O157:H7 infection are similar to those recommended for other foodborne diseases. Basic good food hygiene practice, as described in the WHO "Five keys to safer food", can prevent the transmission of pathogens responsible for many foodborne diseases, and also protect against foodborne diseases caused by STEC.

The five keys to safer food are:

- Keep clean.
- Separate raw and cooked.
- Cook thoroughly.
- Keep food at safe temperatures.
- Use safe water and raw materials.

Such recommendations should in all cases be implemented, especially "cook thoroughly" so that the centre of the food reaches at least 70 °C. Make sure to wash fruits and vegetables carefully, especially if they are eaten raw. If possible, vegetables and fruits should be peeled. Vulnerable populations (such as small children and the elderly) should avoid the consumption of raw or undercooked meat products, raw milk, and products made from raw milk.

Regular hand washing, particularly before food preparation or consumption and after toilet contact, is highly recommended, especially for people who take care of small children, the elderly or immunocompromised individuals, as the bacterium can be passed from person to person, as well as through food, water and direct contact with animals.

A number of STEC infections have been caused by contact with recreational water. Therefore, it is also important to protect such water areas, as well as drinking-water sources, from animal waste.

Campylobacter

Campylobacter (meaning "curved bacteria") is a genus of Gram-negative bacteria. Campylobacter typically appear comma- or s-shaped, and are motile.

Most Campylobacter species can infect humans and other animals, causing disease. The bacterium's main natural reservoir is poultry; humans can contract the disease from eating food contaminated with Campylobacter species. Another source of infection is contact with infected animals, which often carry Campylobacter asymptomatically. At least a dozen species of Campylobacter have been implicated in human disease, with C. jejuni and C. coli being the most common. C. jejuni is now recognized as one of the main causes of bacterial foodborne disease in many developed countries. C. jejuni infection can also spread to the blood in individuals with AIDS, while C. lari is a known cause of recurrent diarrhea in children. C. fetus is a cause of spontaneous abortions in cattle and sheep, as well as an opportunistic pathogen in humans. This genus has been found to be part of the salivary microbiome.

Genetics

The genomes of several Campylobacter species have been sequenced, beginning with C. jejuni in 2000. These genome studies have identified molecular markers specific to members of Campylobacter. Additionally, several markers were found in all Campylobacter species except for C. fetus, the most distantly related species. Many markers were also found which were conserved only between C. jejuni and C. coli, indicating a close relationship between these two species.

Similar studies have investigated the genes responsible for motility in Campylobacter species. All Campylobacter species contain two flagellin genes in tandem for motility, flaA and flaB. These genes undergo intergenic recombination, further contributing to their virulence.

The number of known quinolone-resistant strains is growing. Evidence suggests this is caused by an overuse of this class of antibiotics in animal agriculture.

Bacteriophage

The confusing taxonomy of *Campylobacter* over the past decades makes identifying the earliest reports of *Campylobacter* bacteriophages difficult. Bacteriophages specific to the species now known as *C. coli* and *C. fetus* (previously *Vibrio coli* and *V. fetus*), were first isolated from cattle and pigs during the 1960s, and *Campylobacter* bacteriophage therapy is an ongoing area of research in the age of bacterial antibiotic resistance.

Pathogenesis

Campylobacter can cause a gastrointestinal infection called campylobacteriosis. The incubation period is 24–72 hours after infection. This is characterized by an inflammatory, sometimes bloody diarrhea or dysentery syndrome, mostly including cramps, fever, and pain. The most common routes of transmission are fecal-oral, ingestion of contaminated food or water, and the eating of raw meat. Foods implicated in campylobacteriosis include raw or under-cooked poultry, raw dairy products, and contaminated produce. Campylobacter is sensitive to the stomach's normal production of hydrochloric acid: as a result, the infectious dose is relatively high, and the bacteria rarely cause illness when a person is exposed to less than 10,000 organisms. Nevertheless, people taking antacid medication (e. g. people with gastritis or stomach ulcers) are at higher risk of contracting disease from a smaller number of organisms, since this type of medication neutralizes normal gastric acid.

In humans, the sites of tissue injury include the jejunum, the ileum, and the colon. Most strains of C jejuni produce cytolethal distending toxin, which inhibits cell division and impedes activation of the immune system. This helps the bacteria to evade the immune system and survive for a limited time inside intestinal cells. A cholera-like enterotoxin was also, at one time, believed to be produced, but this appears not to be the case. The organism produces diffuse, bloody, edematous, and exudative enteritis. Campylobacter has, on rare occasions, been blamed for hemolytic uremic syndrome and thrombotic thrombocytopenic purpura, though no unequivocal case reports exist. In some cases, a Campylobacter infection can be the underlying cause of Guillain–Barré syndrome. Gastrointestinal perforation is a rare complication of ileal infection.

Clostridium Perfringens

Clostridium perfringens (formerly known as C. welchii, or Bacillus welchii) is a Gram-positive, rod-shaped, anaerobic, spore-forming pathogenic bacterium of the genus Clostridium. C. perfringens is ever-present in nature and can be found as a normal component of decaying vegetation, marine sediment, the intestinal tract of humans and other vertebrates, insects, and soil. It has the shortest reported generation time of any organism at 6.3 minutes in thioglycolate medium.

C. perfringens is one of the most common causes of food poisoning in the United States, alongside norovirus, Salmonella, Campylobacter, and Staphylococcus aureus. However, it can sometimes be ingested and cause no harm.

Infections due to C. perfringens show evidence of tissue necrosis, bacteremia, emphysematous cholecystitis, and gas gangrene, also known as clostridial myonecrosis. The specific name perfringens is derived from the Latin per (meaning through) and frango (burst), referring to the disruption of tissue that occurs during gas gangrene. The toxin involved in gas gangrene is α-toxin, which inserts into the plasma membrane of cells, producing gaps in the membrane that disrupt normal cellular function. C. perfringens can participate in polymicrobial anaerobic infections. It is commonly encountered in infections as a component of the normal flora. In this case, its role in disease is minor.

The action of C. perfringens on dead bodies is known to mortuary workers as tissue gas. It causes extremely accelerated decomposition, and cannot be stopped by normal embalming measures. These bacteria are resistant to the presence of formaldehyde in normal concentrations.

Genome

C. perfringens has a stable G+C content around 27–28% and average genome size of 3.5 Mb. Genomes of 56 C. perfringens strains have since been made available on NCBI genomes database for the scientific research community. Genomic research has revealed surprisingly high diversity in C. perfringens pangenome, with only 12.6% core genes, identified as the most divergent Gram-positive bacteria reported. Nevertheless, 16S rRNA regions in between C. perfringens strains are found to be highly conserved (sequence identity >99.1%).

Motility

Although they lack flagella, C. perfringens bacteria are able to glide across surfaces because their bodies are lined with filaments from end-to-end. The hypermotile variants such as SM101, are often found arising on the edges of colonies on agar plates. Video microscopy of their gliding movement suggests that they form long, thin filaments that allow them to move rapidly like bacteria with flagella.

Genome sequencing was used to identify the cause(s) of the hypermotile phenotype and their direct derivatives. In comparing them, strains SM124 and SM127, hypermotile derivatives of strains SM101 and SM102, respectively, contained 10 and six nucleotide polymorphisms (SNPs) relative to their parent strains. Mutations in cell division genes is the common feature of the hypermotile strains.

Food Poisoning

Food poisoning in humans is caused by type A strains able to produce the CPE (for Clostridium perfringens enterotoxin). The CPE is a polypeptide of 35.5 kDa that accumulates in the beginning of the sporulation and is excreted to the media when it lysates at the end of the sporulation. It is coded by the cpe gene, present in less than the 5% of the type A strains, and it can be located in the chromosome or in an external plasmid.

In the United Kingdom and United States, C. perfringens bacteria are the third-most common cause of foodborne illness, with poorly prepared meat and poultry, or food properly prepared, but left to stand too long, the main culprits in harboring the bacterium. The C. perfringens enterotoxin (CPE) mediating the disease is heat-labile (inactivated at 74 °C (165 °F)). It can be detected in contaminated food (if not heated properly), and feces. Incubation time is between 6 and 24 (commonly 10–12) hours after ingestion of contaminated food.

Since C. perfringens forms spores that can withstand cooking temperatures, if cooked food is let stand for long enough, germination can ensue and infective bacterial colonies develop. Symptoms typically include abdominal cramping, diarrhea, vomiting, and fever. The whole course usually resolves within 24 hours, but can last up to 2 weeks in older or infirm hosts.

C. perfringens poisoning can also lead to another disease known as enteritis necroticans or Clostridial necrotizing enteritis, (also known as pigbel); this is caused by C. perfringens type C. However, this infection is often fatal. Large numbers of C. perfringens grow in the intestines, and secrete exotoxin. This exotoxin causes necrosis of the intestines, varying levels of hemorrhaging, and perforation of the intestine. Inflammation usually occurs in sections of the jejunum, midsection of the small intestine. This disease eventually leads to septic shock and death. This particular disease is rare in the United States; typically, it occurs in populations with a higher risk. Risk factors for enteritis necroticans include protein-deficient diet, unhygienic food preparation, sporadic feasts of meat (after long periods of a protein-deficient diet), diets containing large amounts of trypsin inhibitors (sweet potatoes), areas prone to infection of the parasite Ascaris (produces a trypsin inhibitor). This disease is contracted in populations living in New Guinea, parts of Africa, Central America, South America, and Asia.

Many cases of C. perfringens food poisoning likely remain subclinical, as antibodies to the toxin are common among the population. This has led to the conclusion that most of the population has experienced food poisoning due to C. perfringens.

Despite its potential dangers, C. perfringens is used as the leavening agent in salt-rising bread. The baking process is thought to reduce the bacterial contamination, precluding negative effects.

Infection

C. perfringens is the most common bacterial agent for gas gangrene. Some symptoms include blisters, tachycardia, swelling, and jaundice.

A strain of C. perfringens might be implicated in multiple sclerosis (MS) nascent (Pattern III) lesions. Tests in mice found that a toxin made by a rare strain of C. perfringens caused MS-like damage in the brain, and earlier work had identified this strain of C. perfringens in a human with MS. MS patients were found to be 10 times more immune-reactive to the epsilon toxin than healthy people.

Prevention

The growth of *C. perfringens* spores can be prevented by most importantly cooking food, especially beef and poultry, thoroughly, to the recommended temperatures. Leftover food should be refrigerated to a temperature below 40 °F (4 °C) within two hours of preparation. Large pots of food such as soup or stew with meats should be divided into small quantities and covered for refrigeration. Leftovers should be reheated to at least 165 °F (74 °C) before serving. A rule of thumb is that if the food tastes, smells, or looks different from what it is supposed to, then the food should be avoided. Even if it looks safe, a food that has been out for a long time can also be dangerous to eat.

Food Poisoning Incidents

On May 7, 2010, "42 residents and 12 staff members at a Louisiana (USA) state psychiatric hospital were affected experienced vomiting, abdominal cramps, and diarrhea". Three patients died within 24 hours. The outbreak was linked to chicken which was cooked a day before it was served and was not cooled down according to hospital guidelines. The outbreak affected 31% of the residents of the hospital and 69% of the staff who ate the chicken. How many of the affected residents ate the chicken is unknown.

In May 2011, a man died after allegedly eating food contaminated with the bacteria on a transatlantic American Airlines flight. The man's wife and daughter are suing American and LSG Sky Chefs, the German company that prepared the inflight food.

In December 2012, a 46-year-old woman died two days after eating a Christmas Day meal at a pub in Hornchurch, Essex, England. She was among about 30 people to fall ill after eating the meal. Samples taken from the victims contained C. perfringens. The hotel manager and the cook were jailed for offences arising from the incident.

In December 2014, 87-year-old Bessie Scott died 3 days after eating a church potluck supper in Nackawic, New Brunswick, Canada. Over 30 other people reported signs of gastrointestinal illness, diarrhea, and abdominal pain. The province's acting chief medical officer says, Clostridium perfringens is the bacteria [sic] that most likely caused the woman's death.

In October 2016, 66-year-old Alex Zdravich died four days after eating an enchilada, burrito, and taco at Agave Azul in West Lafayette, Indiana, United States. Three others who dined the same day reported signs of foodborne illness, which were consistent with the symptoms and rapid onset of C. perfringens infection. They later tested positive for the presence of the bacteria.

In November 2016, food contaminated with C. perfringens caused three individuals to die, and another 22 to be sickened, after a Thanksgiving luncheon hosted by a church in Antioch, California, United States.

In January 2017, a mother and her son sued a restaurant in Rochester, New York, United States as they and 260 other people were sickened after eating foods contaminated with C. perfringens. "Officials from the Monroe County Department of Public Health closed down the Golden Ponds after more than a fourth of its Thanksgiving Day guests became ill. An inspection revealed a walk-in refrigerator with food spills and mold, a damaged gasket preventing the door from closing, and mildew growing inside".

In July 2018, 647 people reported symptoms after eating at a Chipotle Mexican Grill restaurant in Powell, Ohio, United States. Stool samples tested by the CDC tested positive for C. perfringens.

In November 2018, approximately 300 people in Concord, North Carolina, United States were sickened by food at a church barbecue that tested positive for C. perfigens.

Listeria Monocytogenes

Listeria monocytogenes is the species of pathogenic bacteria that causes the infection listeriosis. It is a facultative anaerobic bacterium, capable of surviving in the presence or absence of oxygen. It can grow and reproduce inside the host's cells and is one of the most virulent foodborne pathogens, with 20 to 30% of foodborne listeriosis infections in high-risk individuals may be fatal. Responsible for an estimated 1,600 illnesses and 260 deaths in the United States annually, listeriosis ranks third in total number of deaths among foodborne bacterial pathogens, with fatality rates exceeding even Salmonella spp. and Clostridium botulinum. In the European Union, listeriosis follows an upward trend that began in 2008, causing 2,161 confirmed cases and 210 reported deaths in 2014, 16% more than in 2013. Listeriosis mortality rates are also higher in the EU than for other foodborne pathogens.

Listeria monocytogenes is a Gram-positive bacterium, in the division Firmicutes, named after Joseph Lister. Its ability to grow at temperatures as low as 0 °C permits multiplication at typical refrigeration temperatures, greatly increasing its ability to evade control in human foodstuffs. Motile via flagella at 30 °C and below, but usually not at 37 °C, L. monocytogenes can instead move within eukaryotic cells by explosive polymerization of actin filaments (known as comet tails or actin rockets).

Studies suggest up to 10% of human gastrointestinal tracts may be colonized by L. monocytogenes. Nevertheless, clinical diseases due to L. monocytogenes are more frequently recognized by veterinarians, especially as meningoencephalitis in ruminants.

Due to its frequent pathogenicity, causing meningitis in newborns (acquired transvaginally), pregnant mothers are often advised not to eat soft cheeses such as Brie, Camembert, feta, and queso blanco fresco, which may be contaminated with and permit growth of L. monocytogenes. It is the third-most common cause of meningitis in newborns. Listeria monocytogenes can infect the brain, spinal-cord membranes and the bloodstream of the host through the ingestion of contaminated food such as unpasteurized dairy or raw foods.

Classification

L. monocytogenes is a Gram-positive, non-spore-forming, motile, facultatively anaerobic, rod-shaped bacterium. It is catalase-positive and oxidase-negative, and expresses a beta hemolysin, which causes destruction of red blood cells. This bacterium exhibits characteristic tumbling motility when viewed with light microscopy. Although L. monocytogenes is actively motile by means of peritrichous flagella at room temperature (20–25 °C), the organism does not synthesize flagella at body temperatures (37 °C).

The genus Listeria belongs to the class Bacilli and the order Bacillales, which also includes Bacillus and Staphylococcus. Listeria currently contains 10 species: L. fleischmannii, L. grayi, L. innocua, L. ivanovii, L. marthii, L. monocytogenes, L. rocourtiae, L. seeligeri, L. weihenstephanensis , and L. welshimeri. L. denitrificans, previously thought to be part of the genus Listeria, was reclassified into the new genus Jonesia. Both L. ivanovii and L. monocytogenes are pathogenic in mice, but only L. monocytogenes is consistently associated with human illness. The 13 serotypes of L. monocytogenes can cause disease, but more than 90% of human isolates belong to only three serotypes: 1/2a, 1/2b, and 4b. L. monocytogenes serotype 4b strains are responsible for 33 to 35% of sporadic human cases worldwide and for all major foodborne outbreaks in Europe and North America since the 1980s.

Pathogenesis

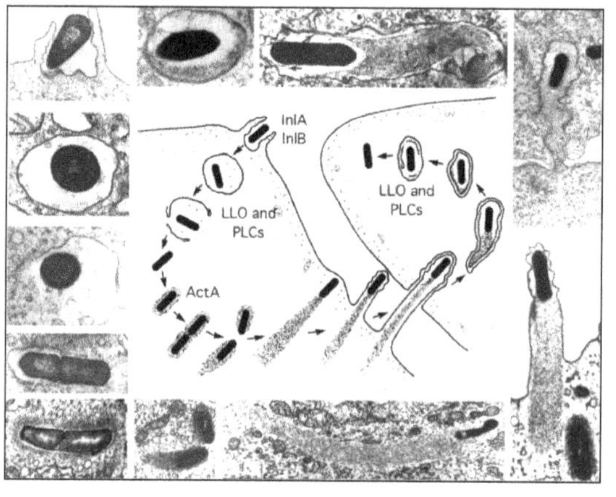

In the figure is shown, stages in the intracellular lifecycle of L. monocytogenes: (Center) Cartoon depicting entry, escape from a vacuole, actin nucleation, actin-based motility, and cell-to-cell spread. (Outside) Representative electron micrographs from which the cartoon was derived. LLO, PLCs, and ActA are all described in the text. The cartoon and micrographs were adapted from Tilney and Portnoy.

Invasive infection by L. monocytogenes causes the disease listeriosis. When the infection is not invasive, any illness as a consequence of infection is termed febrile gastroenteritis. The manifestations of listeriosis include sepsis, meningitis (or meningoencephalitis), encephalitis, corneal ulcer, pneumonia, and intrauterine or cervical infections in pregnant women, which may result in spontaneous abortion (second to third trimester) or stillbirth. Surviving neonates of fetomaternal listeriosis may suffer granulomatosis infantiseptica — pyogenic granulomas distributed over the whole body — and may suffer from physical retardation. Influenza-like symptoms, including persistent fever, usually precede the onset of the aforementioned disorders. Gastrointestinal symptoms, such as nausea, vomiting, and diarrhea, may precede more serious forms of listeriosis or may be the only symptoms expressed. Gastrointestinal symptoms were epidemiologically associated with use of antacids or cimetidine. The onset time to serious forms of listeriosis is unknown, but may range from a few days to 3 weeks. The onset time to gastrointestinal symptoms is unknown, but probably exceeds 12 hours. An early study suggested that L. monocytogenes is unique among Gram-positive bacteria in that it might possess lipopolysaccharide, which serves as an endotoxin. Later, it was found to not be a true endotoxin. Listeria cell walls consistently contain lipoteichoic acids, in which a glycolipid moiety, such as a galactosyl-glucosyl-diglyceride, is covalently linked to the terminal phosphomonoester of the teichoic acid. This lipid region anchors the polymer chain to the cytoplasmic membrane. These lipoteichoic acids resemble the lipopolysaccharides of Gram-negative bacteria in both structure and function, being the only amphipathic polymers at the cell surface.

L. monocytogenes has D-galactose residues on its surface that can attach to D-galactose receptors on the host cell walls. These host cells are generally M cells and Peyer's patches of the intestinal mucosa. Once attached to this cells, L. monocytogenes can translocate past the intestinal membrane and into the body.

The infective dose of L. monocytogenes varies with the strain and with the susceptibility of the victim. From cases contracted through raw or supposedly pasteurized milk, one may safely assume that, in susceptible persons, fewer than 1,000 total organisms may cause disease. L. monocytogenes may invade the gastrointestinal epithelium. Once the bacterium enters the host's monocytes, macrophages, or polymorphonuclear leukocytes, it becomes bloodborne (sepsis) and can grow. Its presence intracellularly in phagocytic cells also permits access to the brain and probably transplacental migration to the fetus in pregnant women. This process is known as the "Trojan Horse mechanism". The pathogenesis of L. monocytogenes centers on its ability to survive and multiply in phagocytic host cells. It seems that Listeria originally evolved to invade membranes of the intestines, as an intracellular infection, and developed a chemical mechanism to do so. This involves a bacterial protein internalin (InlA/InlB), which attaches to a protein on the intestinal cell membrane "cadherin" and allows the bacteria to invade the cells through a zipper mechanism. These adhesion molecules are also to be found in two other unusually tough barriers in humans — the blood-brain barrier and the fetal–placental barrier, and this may explain the apparent affinity that L. monocytogenes has for causing meningitis and affecting babies in utero. Once inside the cell, L. monocytogenes rapidly acidifies the lumen of the vacuole formed around it during cell entry to activate listeriolysin O, a cholesterol-dependent cytolysin capable of disrupting the vacuolar membrane. This frees the pathogen and gives it access to the cytosol of the cell, where it continues its pathogenesis. Motility in the intracellular space is provided by actin assembly-inducing protein, which allows the bacteria to use the host cell's actin polymerization machinery to polymerize the cytoskeleton to give a "boost" to the bacterial cell so it can move in the cell. The same mechanism also allows the bacteria to travel from cell to cell.

Regulation of Pathogenesis

L. monocytogenes can act as a saprophyte or a pathogen, depending on its environment. When this bacterium is present within a host organism, quorum sensing and other signals cause the up-regulation of several virulence genes. Depending on the location of the bacterium within the host organism, different activators up-regulate the virulence genes. SigB, an alternative sigma factor, up-regulates Vir genes in the intestines, whereas PrfA up-regulates gene expression when the bacterium is present in blood. L. monocytogenes also senses the etry to host by examining available nutrient sources. For example L-glutamine, an abundant nitrogen source in the host, induces the expression of virulence genes in L. monocytogenes. Little is known about how this bacterium switches between acting as a saprophyte and a pathogen; however, several noncoding RNAs are thought to be required to induce this change.

Pathogenicity of Lineages

L. monocytogenes has three distinct lineages, with differing evolutionary histories and pathogenic potentials. Lineage I strains contain the majority of human clinical isolates and all human epidemic clones, but are underrepresented in animal clinical isolates. Lineage II strains are overrepresented in animal cases and underrepresented in human clinical cases, and are more prevalent in environmental and food samples. Lineage III isolates are very rare, but significantly more common in animal than human isolates.

Detection

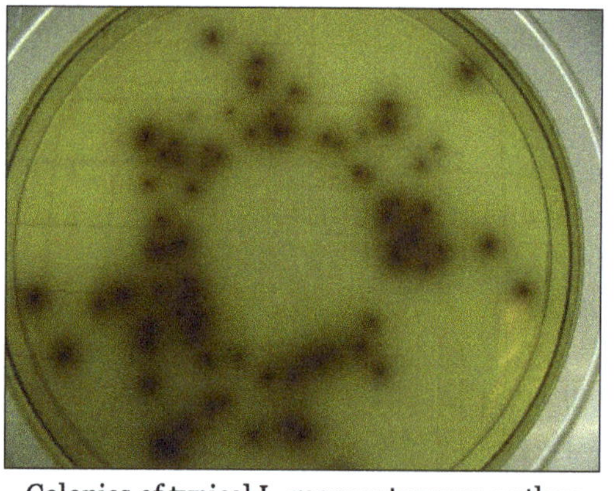

Colonies of typical L. monocytogenes as they appear when grown on Listeria-selective agar.

The Anton test is used in the identification of L. monocytogenes; instillation of a culture into the conjunctival sac of a rabbit or guinea pig causes severe keratoconjunctivitis within 24 hours.

Listeria species grow on media such as Mueller-Hinton agar. Identification is enhanced if the primary cultures are done on agar containing sheep blood, because the characteristic small zone of hemolysis can be observed around and under colonies. Isolation can be enhanced if the tissue is kept at 4 °C for some days before inoculation into bacteriologic media. The organism is a facultative anaerobe and is catalase-positive and motile. *Listeria* produces acid, but not gas, in a variety of carbohydrates. The motility at room temperature and hemolysin production are primary findings that help differentiate listeria from corynebacterium.

The methods for analysis of food are complex and time-consuming. The present U.S. FDA method, revised in September 1990, requires 24 and 48 hours of enrichment, followed by a variety of other tests. Total time to identification takes five to seven days, but the announcement of specific nonradiolabelled DNA probes should soon allow a simpler and faster confirmation of suspect isolates.

Recombinant DNA technology may even permit two- to three-day positive analysis in the future. Currently, the FDA is collaborating in adapting its methodology to quantitate very low numbers of the organisms in foods.

Use as a Transfection Vector

Because L. monocytogenes is an intracellular bacterium, some studies have used this bacterium as a vector to deliver genes in vitro. Current transfection efficiency remains poor. One example of the successful use of L. monocytogenes in in vitro transfer technologies is in the delivery of gene therapies for cystic fibrosis cases.

Epidemiology

Researchers have found Listeria monocytogenes in at least 37 mammalian species, both domesticated and feral, as well as in at least 17 species of birds and possibly in some species of fish and shellfish. Laboratories can isolate Listeria monocytogenes from soil, silage, and other environmental sources. Listeria monocytogenes is quite hardy and resists the deleterious effects of freezing, drying, and heat remarkably well for a bacterium that does not form spores. Most Listeria monocytogenes strains are pathogenic to some degree.

Routes of Infection

Listeria monocytogenes has been associated with such foods as raw milk, pasteurized fluid milk, cheeses (particularly soft-ripened varieties), ice cream, raw vegetables, fermented raw-meat sausages, raw and cooked poultry, raw meats (of all types), and raw and smoked fish. Most bacteria can survive near freezing temperatures, but cannot absorb nutrients, grow or replicate; however, L. monocytogenes has the ability to grow at temperatures as low as 0 °C which permits exponential multiplication in refrigerated foods. At refrigeration temperature, such as 4 °C, the amount of ferric iron can affect the growth of L. monocytogenes.

Infectious Cycle

The primary site of infection is the intestinal epithelium, where the bacteria invade nonphagocytic cells via the "zipper" mechanism. Uptake is stimulated by the binding of listerial internalins (Inl) to E-cadherin, a host cell adhesion factor, or Met (c-Met), hepatocyte growth factor. This binding activates certain Rho-GTPases, which subsequently bind and stabilize Wiskott Aldrich syndrome protein (WAsp). WAsp can then bind the Arp2/3 complex and serve as an actin nucleation point. Subsequent actin polymerization creates a "phagocytic cup", an actin-based structure normally formed around foreign materials by phagocytes prior to endocytosis. The net effect of internalin binding is to exploit the junction-forming apparatus of the host into internalizing the bacterium. L. monocytogenes can also invade phagocytic cells (e.g., macrophages), but requires only internalins for invasion of nonphagocytic cells.

Following internalization, the bacterium must escape from the vacuole/phagosome before fusion with a lysosome can occur. Three main virulence factors that allow the bacterium to escape are listeriolysin O (LLO-encoded by hly) phospholipase A (encoded by plcA) and phospholipase B (plcB). Secretion of LLO and PlcA disrupts the vacuolar membrane and allows the bacterium to escape into the cytoplasm, where it may proliferate.

Once in the cytoplasm, L. monocytogenes exploits host actin for the second time. ActA proteins associated with the old bacterial cell pole (being a bacillus, L. monocytogenes septates in the middle of the cell, thus has one new pole and one old pole) are capable of binding the Arp2/3 complex, thereby inducing actin nucleation at a specific area of the bacterial cell surface. Actin polymerization then propels the bacterium unidirectionally into the host cell membrane. The protrusion formed may then be internalized by a neighboring cell, forming a double-membrane vacuole from which the bacterium must escape using LLO and PlcB. This mode of direct cell-to-cell spread involves a cellular mechanism known as paracytophagy.

Staphylococcus Aureus

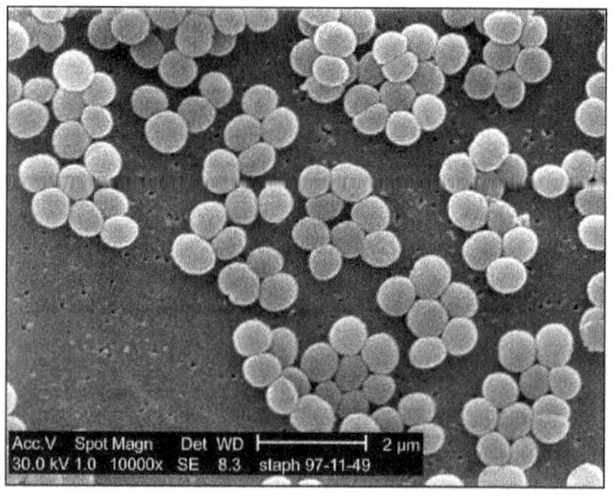

Staphylococcus aureus.

Staphylococcal enteritis is an inflammation that is usually caused by eating or drinking substances contaminated with staph enterotoxin. The toxin, not the bacterium, settles in the small intestine and causes inflammation and swelling. This in turn can cause abdominal pain, cramping, dehydration, diarrhea and fever.

Staphylococcus aureus is a Gram-positive, facultative anaerobe, coccal (round shaped) bacteria that appears in grape-like clusters that can thrive in high salt and low water activity habitats. S. aureus bacteria can live on the skin which is one of the primary modes of transmission. S. aureus can cause a range of illnesses from minor skin infections to Staphylococcus aureus food poisoning enteritis. Since humans are the primary source, cross-contamination is the most common way the microorganism is introduced into foods. Foods at high risks are those prepared in large quantities. Staphylococcus aureus is a true food poisoning organism. It produces a heat stable enterotoxin when allowed to grow for several hours in foods such as cream-filled baked goods, poultry meat, gravies, eggs, meat salads, puddings and vegetables. It is important to note that the toxins may be present in dangerous amounts in foods that have no signs of spoilage, such as a bad smell, any off color, odor, or textural or flavor change.

Enteritis is the inflammation of the small intestine. It is generally caused by eating or drinking substances that are contaminated with bacteria or viruses. The bacterium and toxin settles in the small intestine and cause inflammation and swelling. This in turn can cause abdominal pain, cramping, diarrhea, fever, and dehydration. There are other types of enteritis, the types include: bacterial gastroenteritis, Campylobacter enteritis, E. coli enteritis, radiation enteritis, Salmonella enteritis and Shigella enteritis.

Symptoms

Common symptoms of Staphylococcus aureus food poisoning include: a rapid onset which is usually 1–6 hours, nausea, explosive vomiting for up to 24 hours, abdominal cramps/pain, headache, weakness, diarrhea and usually a subnormal body temperature. Symptoms usually start one to six hours after eating and last less than 12 hours. The duration of some cases may take two or more days to fully resolve.

Pathogenesis

S. aureus is an enterotoxin producer. Enterotoxins are chromosomally encoded exotoxins that are produced and secreted from several bacterial organisms. It is a heat stable toxin and is resistant to digestive protease. It is the ingestion of the toxin that causes the inflammation and swelling of the intestine.

Prevention

Staphylococcal enteritis may be avoided by using proper hygiene and sanitation with food preparation. This includes thoroughly cooking all meats. If food is to be stored longer than two hours, keep hot foods hot (over 140 °F) and cold foods cold (40 °F or under). Ensure to refrigerate leftovers promptly and store cooked food in a wide, shallow container and refrigerate as soon as possible. Sanitation is very important. Keep kitchens and food-serving areas clean and sanitized. Finally, as most staphylococcal food poisoning are the result of food handling, hand washing is critical. Food handlers should use hand sanitizers with alcohol or thorough hand washing with soap and water.

Tips for hand washing:

- Wash hands with warm, soapy water before and after handling raw foods:
  - First, wet your hands.
  - Add soap to your hands.
  - Rub both sides for at least 20 seconds.
  - Rinse thoroughly.
  - Air dry, or dry your hands with a clean towel or paper towel.
- Always wash your hands after using the bathroom, after changing a baby's diaper, after touching pets or other animals, and after sneezing or coughing.
- Properly dress or glove.

Clostridium Botulinum

Clostridium botulinum is a Gram-positive, rod-shaped, anaerobic, spore-forming, motile bacterium with the ability to produce the neurotoxin botulinum.

The botulinum toxin can cause a severe flaccid paralytic disease in humans and other animals and is the most potent toxin known to mankind, natural or synthetic, with a lethal dose of 1.3–2.1 ng/kg in humans.

C. botulinum is a diverse group of pathogenic bacteria initially grouped together by their ability to produce botulinum toxin and now known as four distinct groups, C. botulinum groups I–IV. C. botulinum groups I–IV, as well as some strains of Clostridium butyricum and Clostridium baratii, are the bacteria responsible for producing botulinum toxin.

C. botulinum is responsible for foodborne botulism (ingestion of preformed toxin), infant botulism (intestinal infection with toxin-forming C. botulinum), and wound botulism (infection of a wound with C. botulinum). C. botulinum produces heat-resistant endospores that are commonly found in soil and are able to survive under adverse conditions.

C. botulinum is commonly associated with bulging canned food; bulging, misshapen cans are due to an internal increase in pressure caused by gas produced by the bacteria.

Microbiology

C. botulinum is a Gram-positive, rod-shaped, spore-forming bacterium. It is an obligate anaerobe, meaning that oxygen is poisonous to the cells. However, C. botulinum tolerates traces of oxygen due to the enzyme superoxide dismutase, which is an important antioxidant defense in nearly all cells exposed to oxygen. C. botulinum is only able to produce the neurotoxin during sporulation, which can only happen in an anaerobic environment. Other bacterial species produce spores in an unfavorable growth environment to preserve the organism's viability and permit survival in a dormant state until the spores are exposed to favorable conditions.

C. botulinum is divided into four distinct phenotypic groups (I-IV) and is also classified into seven serotypes (A-G) based on the antigenicity of the botulinum toxin produced.

Groups

The classification into groups is based on the ability of the organism to digest complex proteins. Studies at the DNA and rRNA level support the subdivision of the species into groups I-IV. Most outbreaks of human botulism are caused by group I (proteolytic) or II (non-proteolytic) *C. botulinum*. Group III organisms mainly cause diseases in animals. Group IV *C. botulinum* has not been shown to cause human or animal disease.

Botulinum Toxin

Neurotoxin production is the unifying feature of the species. Eight types of toxins have been identified that are allocated a letter (A–H), several of which can cause disease in humans. They are resistant to degradation by enzymes found in the gastrointestinal tract. This allows for ingested toxin to be absorbed from the intestines into the bloodstream. However, all types of botulinum toxin are rapidly destroyed by heating to 100 °C for 15 minutes (900 seconds).

Most strains produce one type of neurotoxin, but strains producing multiple toxins have been described. C. botulinum producing B and F toxin types have been isolated from human botulism cases in New Mexico and California. The toxin type has been designated Bf as the type B toxin was found in excess to the type F. Similarly, strains producing Ab and Af toxins have been reported. Evidence indicates the neurotoxin genes have been the subject of horizontal gene transfer, possibly from a viral source. This theory is supported by the presence of integration sites flanking the toxin in some strains of C. botulinum. However, these integrations sites are degraded, indicating that the C. botulinum acquired the toxin genes quite far in the evolutionary past.

Botulinum Toxin Types

Only botulinum toxin types A, B, E, F and H cause disease in humans. Types A, B, and E are associated with food-borne illness, with type E specifically associated with fish products. Type C produces limber-neck in birds and type D causes botulism in other mammals. No disease is associated with type G. The "gold standard" for determining toxin type is a mouse bioassay, but the genes for types A, B, E, and F can now be readily differentiated using quantitative PCR. As no antitoxin to type H is yet available, discovered in 2013 and by far the deadliest, details are kept under shroud.

A few strains from organisms genetically identified as other Clostridium species have caused human botulism: C. butyricum has produced type E toxin and C. baratii had produced type F toxin. The ability of C. botulinum to naturally transfer neurotoxin genes to other clostridia is concerning, especially in the food industry, where preservation systems are designed to destroy or inhibit only C. botulinum but not other Clostridium species.

Phenotypic groups of Clostridium botulinum				
Properties	Group I	Group II	Group III	Group IV
Toxin Types	A, B, F	B, E, F	C, D	G
Proteolysis	+	–	weak	–
Saccharolysis	–	+	–	–
Disease host	human	human	animal	–
Toxin gene	chromosome/plasmid	chromosome/plasmid	bacteriophage	plasmid
Close relatives	C. sporogenes C. putrificum	C. butyricum C. beijerinickii	C. haemolyticum C. novyi type A	C. subterminale C. haemolyticum

Laboratory Isolation

In the laboratory, *C. botulinum* is usually isolated in tryptose sulfite cycloserine (TSC) growth medium in an anaerobic environment with less than 2% oxygen. This can be achieved by several commercial kits that use a chemical reaction to replace O_2 with CO_2. *C. botulinum* is a lipase-positive microorganism that grows between pH of 4.8 and 7.0 and cannot use lactose as a primary carbon source, characteristics important for biochemical identification.

Pathology

Botulism poisoning can occur due to preserved or home-canned, low-acid food that was not processed using correct preservation times and pressure.

Foodborne botulism "Signs and symptoms of foodborne botulism typically begin between 18 and 36 hours after the toxin gets into your body, but can range from a few hours to several days, depending on the amount of toxin ingested."

- Double vision,
- Blurred vision,

- Drooping eyelid,
- Nausea, vomiting, and abdominal cramps,
- Slurred speech,
- Trouble breathing,
- Difficulty in swallowing,
- Dry mouth,
- Muscle weakness,
- Constipation,
- Reduced or absent deep tendon reactions, such as in the knee.

Wound botulism Most people who develop wound botulism inject drugs several times a day, so it's difficult to determine how long it takes for signs and symptoms to develop after the toxin enters the body. Most common in people who inject black tar heroin, wound botulism signs and symptoms include:

- Difficulty swallowing or speaking,
- Facial weakness on both sides of the face,
- Blurred or double vision,
- Drooping eyelids,
- Trouble breathing,
- Paralysis.

Infant botulism If infant botulism is related to food, such as honey, problems generally begin within 18 to 36 hours after the toxin enters the baby's body. Signs and symptoms include:

- Constipation (often the first sign),
- Floppy movements due to muscle weakness and trouble controlling the head,
- Weak cry,
- Irritability,
- Drooling,
- Drooping eyelids,
- Tiredness,
- Difficulty sucking or feeding,
- Paralysis.

Beneficial effects of botulinum toxin: Purified botulinum toxin is diluted by a physician for treatment:

- Congenital pelvic tilt,
- Spasmodic dysphasia (the inability of the muscles of the larynx),
- Achalasia (esophageal stricture),
- Strabismus (crossed eyes),
- Paralysis of the facial muscles,
- Failure of the cervix,
- Blinking frequently,
- Anti-cancer drug delivery.

Adult intestinal toxemia: A very rare form of botulism that occurs by the same route as infant botulism but is among adults. Occurs rarely and sporadically. Signs and symptoms include:

- Abdominal pain,
- Blurred vision,
- Diarrhea,
- Dysarthria,
- Imbalance,
- Weakness in arms and hand area.

Use and Detection

C. botulinum is used to prepare the medicaments Botox, Dysport, Xeomin, and Neurobloc used to selectively paralyze muscles to temporarily relieve muscle function. It has other "off-label" medical purposes, such as treating severe facial pain, such as that caused by trigeminal neuralgia.

Botulinum toxin produced by C. botulinum is often believed to be a potential bioweapon as it is so potent that it takes about 75 nanograms to kill a person (LD50 of 1 ng/kg, assuming an average person weighs ~75 kg); 1 kilogram of it would be enough to kill the entire human population. For comparative purposes, a quarter of a typical grain of sand's weight (350 ng) of botulinum toxin would constitute a lethal dose for humans.

A "mouse protection" or "mouse bioassay" test determines the type of C. botulinum toxin present using monoclonal antibodies. An enzyme-linked immunosorbent assay (ELISA) with digoxigen-in-labeled antibodies can also be used to detect the toxin, and quantitative PCR can detect the toxin genes in the organism.

Growth Conditions and Prevention

C. botulinum is a soil bacterium. The spores can survive in most environments and are very hard to kill. They can survive the temperature of boiling water at sea level, thus many foods are canned with a pressurized boil that achieves even higher temperatures, sufficient to kill the spores.

C. botulinum is an obligate anaerobe that is widely distributed in nature and is assumed to be present on all food surfaces. Its optimum growth temperature is within the mesophilic range. In spore form, it is the most heat resistant pathogen that can survive in low acid foods and grow to produce toxin. The toxin attacks the nervous system and will kill an adult at a dose of around 75 ng. This toxin is detoxified by holding food at 100 °C for 10 minutes.

Growth of the bacterium can be prevented by high acidity, high ratio of dissolved sugar, high levels of oxygen, very low levels of moisture, or storage at temperatures below 3 °C (38 °F) for type A. For example, in a low-acid, canned vegetable such as green beans that are not heated enough to kill the spores (i.e., a pressurized environment) may provide an oxygen-free medium for the spores to grow and produce the toxin. However, pickles are sufficiently acidic to prevent growth; even if the spores are present, they pose no danger to the consumer. Honey, corn syrup, and other sweeteners may contain spores, but the spores cannot grow in a highly concentrated sugar solution; however, when a sweetener is diluted in the low-oxygen, low-acid digestive system of an infant, the spores can grow and produce toxin. As soon as infants begin eating solid food, the digestive juices become too acidic for the bacterium to grow.

The control of food-borne botulism caused by C. botulinum is based almost entirely on thermal destruction (heating) of the spores or inhibiting spore germination into bacteria and allowing cells to grow and produce toxins in foods. Conditions conducive of growth are dependent on various environmental factors. Growth of C. botulinum is a risk in low acid foods as defined by having a pH below 4.6 although growth is significantly retarded for pH below 4.9. There have been some cases and specific conditions reported to sustain growth with pH below 4.6.

Vibrio Vulnificus

Vibrio vulnificus is a species of Gram-negative, motile, curved rod-shaped (bacillus), pathogenic bacteria of the genus Vibrio. Present in marine environments such as estuaries, brackish ponds, or coastal areas, V. vulnificus is related to V. cholerae, the causative agent of cholera.

Infection with V. vulnificus leads to rapidly expanding cellulitis or sepsis. It was first isolated as a source of disease in 1976.

Signs and Symptoms

V. vulnificus is an extremely virulent bacterium that can cause three types of infections:

- Acute gastroenteritis from eating raw or undercooked shellfish: V. vulnificus causes an infection often incurred after eating seafood, especially raw or undercooked oysters. It does not alter the appearance, taste, or odor of oysters. Symptoms include vomiting, diarrhea, and abdominal pain.
- Necrotizing wound infections can occur in injured skin exposed to contaminated marine water. V. vulnificus bacteria can enter the body through open wounds when swimming or wading in infected waters, or by puncture wounds from the spines of fishes such as stingrays. People may develop a blistering dermatitis sometimes mistaken for pemphigus or pemphigoid.

- Invasive sepsis can occur after eating raw or undercooked shellfish, especially oysters. V. vulnificus is 80 times more likely to spread into the bloodstream in people with compromised immune systems, especially those with chronic liver disease. When this happens, severe symptoms including blistering skin lesions and septic shock can sometimes lead to death. This severe infection may occur regardless of whether the infection began from contaminated food or an open wound.

Among healthy people, ingestion of V. vulnificus can cause vomiting, diarrhea, and abdominal pain. In someone with a compromised immune system, particularly those with chronic liver disease, it can infect the bloodstream, causing a severe and life-threatening illness characterized by fever and chills, decreased blood pressure (septic shock), and blistering skin lesions. While men have been shown to be more at risk from this infection than women, co-morbidities such as alcoholic cirrhosis and diseases affecting the endocrine system (diabetes, rheumatoid arthritis, etc.) put a person far more at risk to develop infection from V. vulnificus.

Pathogenesis

Capsule: V. *vulnificus* has a capsule, made of polysaccharides, and is thought to protect against phagocytosis. The capsule also aids the bacteria in escaping opsonization. Different strains of the bacteria are capable of shifting through the unencapsulated and encapsulated forms. Mouse models have shown that the unencapsulated forms are avirulent. These same strains however, are shown to have a higher predisposition to shift to the virulent encapsulated form when taken up by oysters.

Endotoxin: Like all gram negative bacteria, V. vulnificus has LPS (lipopolysaccharide as the major component of its outer membrane). However, the LPS the bacteria produces isn't as efficient at triggering the immune system's release of tumor necrosis factor (TNF) alpha and other cytokines that produce shock syndromes. The capsular proteins the bacteria express however, are capable of producing an immune response contributing to shock syndrome.

Exotoxin: V. vulnificus produces a number of extracellular toxins such as metalloprotease VvpE, cytolysin/hemolysin VvhA, and the multifunctional autoprocessing repeats-in-toxins (MARTX) toxin. While the VvhA and MARTX toxin are factors in the bacteria's virulence, in vivo studies in mice suggest that the MARTX toxin is more responsible for bacterial dissemination from the intestine to produce sepsis.

Iron: Growth of V. vulnificus is dependent on the amount of iron that is accessible to the bacteria. The observed association of the infection with liver disease (associated with increased serum iron) might be due to the capability of more virulent strains to capture iron bound to transferrin.

Strains

The most harmful strains of V. vulnificus documented have been observed in three different forms. The first is in an anti-phagocytic polysaccharide capsule that protects the bacteria. By encapsulating the bacteria, phagocytosis and opsonization are not able to occur, thus allowing the bacteria to continue throughout the organism it is in. The second way that V. vulnificus has been most harmful is with some of the toxins that it creates. These toxins are not part of the infection that V. vulnificus causes but instead they are part of a secondary infection in the GI tract that most certainly will lead to systemic infection. Lastly, V. vulnificus has been seen to cause more harm in patients that have higher levels of iron.

Prognosis

V. vulnificus may not be a commonly known bacteria, but it is, however, the most common cause of death due to seafood in the United States. Infection and mortality due to V. vulnificus causes over 95% of deaths in the United States that are known to have happened because of ingested seafood. Surprisingly enough, while V. vulnificus claims 95% of seafood related deaths, if treatment with tetracycline or other cephalosporin antibiotics is initiated at the onset of symptoms and is treated appropriately, the patient will experience no long term effects provided they continue to take the full course over antibiotics which is typically about two weeks.

The worst prognosis is in those people arriving at hospital in a state of shock. Total mortality in treated people (ingestion and wound) is around 33%.

People especially vulnerable are those with liver disease (especially cirrhosis and hepatitis) or immunocompromised states (some kinds of cancer, bone marrow suppression, HIV, diabetes, etc). With these cases, V. vulnificus usually enters the bloodstream, where it may cause fever and chills, septic shock (with sharply decreased blood pressure), and blistering skin lesions. About half of those who contract blood infections die.

V. vulnificus infections also disproportionately affect males; 85% of those developing endotoxic shock from the bacteria are male. Females having had an oophorectomy experienced increased mortality rates, as estrogen has been shown experimentally to have a protective effect against V. vulnificus.

Norovirus

Norovirus is a virus that's very contagious. It passes easily through direct or indirect contact with an infected person. It can spread quickly in close quarters such as hospitals, schools, and day care centers.

Most people have some experience with norovirus. It's a common illness of the stomach and intestinal tract. Norovirus is a source of food poisoning because you can get it from eating contaminated food. The results are the same no matter how you get it.

The hallmark symptoms of norovirus are vomiting and watery, non-bloody diarrhea. These symptoms usually start within 12 to 48 hours of being exposed and can last up to three days. Most people make a full recovery.

There's no specific treatment except to rest and rehydrate. The most significant complication is dehydration. Norovirus can be serious and even fatal in the very young, older adults, and people with other health problems.

Because there are many norovirus strains, having it once doesn't prevent you from getting it again. You can lower the risk of disease transmission by washing your hands thoroughly and frequently.

Norovirus is thought to be the most common cause of acute gastroenteritis in the world, leading to 685 million cases a year. Gastroenteritis is inflammation and infection of the stomach and intestinal tract caused by any infectious organism, like bacteria and viruses. Norovirus is responsible for up to 21 million illnesses in the United States each year.

Norovirus Symptoms

Symptoms of infection usually start somewhere between 12 and 48 hours after you've been exposed to the virus. They can range from quite mild to severe. Some signs and symptoms of norovirus are:

- Nausea and vomiting,
- Abdominal cramping or pain,
- Watery stools or diarrhea,
- Low-grade fever,
- Chills,
- Headache,
- Generalized body aches.

Symptoms usually last between 24 to 72 hours. See your doctor if symptoms persist beyond that or if you see blood in your stools. Severe diarrhea can lead to dehydration, which should be considered a medical emergency. Signs and symptoms of dehydration include:

- Dry mouth and throat,
- Decreased output of urine or dark urine,
- No wet diaper for 6 to 8 hours in infants,
- No urine in 12 hours for children,
- Sunken eyes,
- Sleepiness and fatigue,
- Headache,
- Dizziness,
- Confusion and lethargy,
- Rapid heart rate.

Dehydration can be life-threatening, especially for the following groups:

- People with a weakened immune system.
- People with preexisting health conditions.
- The very old and the very young.
- Organ or stem cell transplant recipients.

Campylobacteriosis

Campylobacteriosis is an infection by the Campylobacter bacterium, most commonly C. jejuni. It is among the most common bacterial infections of humans, often a foodborne illness. It produces an inflammatory, sometimes bloody, diarrhea or dysentery syndrome, mostly including cramps, fever and pain.

Symptoms

The prodromal symptoms are fever, headache, and myalgia, which can be severe, lasting as long as 24 hours. After 1–5 days, typically, these are followed by diarrhea (as many as 10 watery, frequently bloody, bowel movements per day) or dysentery, cramps, abdominal pain, and fever as high as 40 °C (104 °F). In most people, the illness lasts for 2–10 days. It is classified as invasive/ inflammatory diarrhea, also described as bloody diarrhea or dysentery.

There are other diseases showing similar symptoms. For instance, abdominal pain and tenderness may be very localized, mimicking acute appendicitis. Furthermore, *Helicobacter pylori* is closely related to Campylobacter and causes peptic ulcer disease.

Complications

Complications include toxic megacolon, dehydration and sepsis. Such complications generally occur in young children (< 1 year of age) and immunocompromised people. A chronic course of the disease is possible; this disease process is likely to develop without a distinct acute phase. Chronic campylobacteriosis features a long period of sub-febrile temperature and asthenia; eye damage, arthritis, endocarditis may develop if infection is untreated.

Occasional deaths occur in young, previously healthy individuals because of blood volume depletion (due to dehydration), and in persons who are elderly or immunocompromised.

Some individuals (1–2 in 100,000 cases) develop Guillain–Barré syndrome, in which the nerves that join the spinal cord and brain to the rest of the body are damaged, sometimes permanently. This occurs only with infection of *C. jejuni* and *C. upsaliensis*.

Other Factors

In patients with HIV, infections may be more frequent, may cause prolonged bouts of dirty brown diarrhea, and may be more commonly associated with bacteremia and antibiotic resistance. In participants of unprotected anal intercourse, campylobacteriosis is more localized to the distal end of the colon and may be termed a proctocolitis. The severity and persistence of infection in patients with AIDS and hypogammaglobulinemia indicates that both cell-mediated and humoral immunity are important in preventing and terminating infection.

Cause

Campylobacteriosis is caused by Campylobacter bacteria (curved or spiral, motile, non–spore-forming, Gram-negative rods). The disease is usually caused by C. jejuni, a spiral and comma shaped bacterium normally found in cattle, swine, and birds, where it is nonpathogenic, but the illness can also be caused by C. coli (also found in cattle, swine, and birds), C. upsaliensis (found in cats and dogs) and C. lari (present in seabirds in particular).

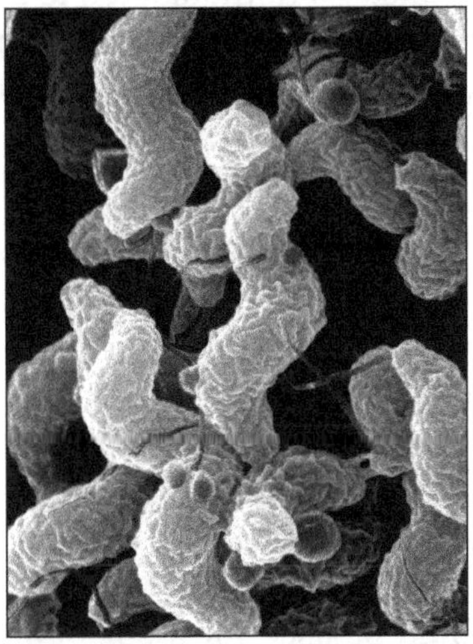

Campylobacter bacteria are the number-one cause of food-related gastrointestinal illness in the United States. This scanning electron microscope image shows the characteristic spiral, or corkscrew, shape of *C. jejuni* cells and related structures.

One effect of campylobacteriosis is tissue injury in the gut. The sites of tissue injury include the jejunum, the ileum, and the colon. C jejuni appears to achieve this by invading and destroying epithelial cells.

C. jejuni can also cause a latent autoimmune effect on the nerves of the legs, which is usually seen several weeks after a surgical procedure of the abdomen. The effect is known as an acute idiopathic demyelinating polyneuropathy (AIDP), i.e. Guillain–Barré syndrome, in which one sees symptoms of ascending paralysis, dysaesthesias usually below the waist, and, in the later stages, respiratory failure.

Some strains of C jejuni produce a cholera-like enterotoxin, which is important in the watery diarrhea observed in infections. The organism produces diffuse, bloody, edematous, and exudative enteritis. In a small number of cases, the infection may be associated with hemolytic uremic syndrome and thrombotic thrombocytopenic purpura through a poorly understood mechanism.

Transmission

The common routes of transmission for the disease-causing bacteria are fecal-oral, person-to-person sexual contact, ingestion of contaminated food (generally unpasteurized (raw) milk and undercooked or poorly handled poultry), and waterborne (i.e., through contaminated drinking water). Contact with contaminated poultry, livestock, or household pets, especially puppies, can also cause disease.

Animals farmed for meat are the main source of campylobacteriosis. A study published in PLoS Genetics by researchers from Lancashire, England, and Chicago, Illinois, found that 97 percent of campylobacteriosis cases sampled in Lancashire were caused by bacteria typically found in chicken and livestock. In 57 percent of cases, the bacteria could be traced to chicken, and in 35 percent to cattle. Wild animal and environmental sources were accountable for just three percent of disease.

The infectious dose is 1000–10,000 bacteria (although ten to five hundred bacteria can be enough to infect humans). Campylobacter species are sensitive to hydrochloric acid in the stomach, and acid reduction treatment can reduce the amount of inoculum needed to cause disease.

Exposure to bacteria is often more common during travelling, and therefore campylobacteriosis is a common form of travelers' diarrhea.

Prevention

- Pasteurization of milk and chlorination of drinking water destroys the organisms.
- Treatment with antibiotics can reduce fecal excretion.
- Infected health care workers should not provide direct patient care.
- Separate cutting boards should be used for foods of animal origin and other foods. After preparing raw food of animal origin, all cutting boards and countertops should be carefully cleaned with soap and hot water.
- Contact with pet saliva and feces should be avoided.

The World Health Organization recommends the following:

- Food should be properly cooked and hot when served.
- Consume only pasteurized or boiled milk and milk products, never raw milk products.
- Make sure that ice is from safe water.
- If you are not sure of the safety of drinking water, boil it, or disinfect it with chemical disinfectant.
- Wash hands thoroughly and frequently with soap, especially after using the toilet and after contact with pets and farm animals.
- Wash fruits and vegetables thoroughly, especially if they are to be eaten raw. Peel fruits and vegetables whenever possible.
- Food handlers, professionals and at home, should observe hygienic rules during food preparation.
- Professional food handlers should immediately report to their employer any fever, diarrhea, vomiting or visible infected skin lesions.

Prognosis

Campylobacteriosis is usually self-limited without any mortality (assuming proper hydration is maintained). However, there are several possible complications.

Giardiasis

Giardiasis is an infection in your small intestine. It's caused by a microscopic parasite called Giardia lamblia. Giardiasis spreads through contact with infected people. And you can get giardiasis by eating contaminated food or drinking contaminated water. Pet dogs and cats also frequently contract giardia.

This condition can be found all over the world, according to the Centers for Disease Control and Prevention (CDC). However, it's more common in overcrowded developing countries that lack sanitary conditions and water quality control.

What are the Causes of Giardiasis

G. lamblia are found in animal and human feces. These parasites also thrive in contaminated food, water, and soil, and can survive outside a host for long periods of time. Accidentally consuming these parasites can lead to an infection.

The most common way to get giardiasis is to drink water that contain G. lamblia. Contaminated water can be in swimming pools, spas, and bodies of water, such as lakes. Sources of contamination include animal feces, diapers, and agricultural runoff.

Contracting giardiasis from food is less common because heat kills the parasites. Poor hygiene when handling food or eating produce rinsed in contaminated water can allow the parasite to spread.

Giardiasis also spreads through personal contact. For example, unprotected anal sex can pass the infection from one person to another.

Changing a child's diaper or picking up the parasite while working in a day care center are also common ways to become infected. Children are at high risk for giardiasis because they're likely to encounter feces when wearing diapers or potty training.

What are the Symptoms of Giardiasis

Some people can carry giardia parasites without experiencing any symptoms. Symptoms of giardiasis generally show up one or two weeks after exposure. Common symptoms include:

- Fatigue,
- Nausea,
- Diarrhea or greasy stools,
- Loss of appetite,
- Vomiting,
- Bloating and abdominal cramps,
- Weight loss,

- Excessive gas,
- Headaches,
- Abdominal pain.

Toxoplasmosis

Toxoplasmosis is an infection caused by a parasite. This parasite is called Toxoplasma gondii. It can be found in cat feces and undercooked meat, especially venison, lamb, and pork. It can also be transmitted through contaminated water. Toxoplasmosis can be deadly or cause serious birth defects for a fetus if the mother becomes infected. This is why doctors recommend against pregnant woman scooping or cleaning cat litter boxes.

Most people who have toxoplasmosis never have any symptoms at all. According to the Centers for Disease Control and Prevention (CDC), over 60 million people in the United States are infected with the parasite. The people who are most at risk for serious infections are those with compromised immune systems and infants born to mothers with active infection during their pregnancy.

What are the Symptoms of Toxoplasmosis

Most people who've been infected with the parasite that causes toxoplasmosis show no signs or symptoms.

People who develop symptoms may experience:

- A fever,
- Swollen lymph nodes, especially in the neck,
- A headache,
- Muscle aches and pains,
- Sore throat.

These symptoms can last for a month or more and usually resolve on their own.

Toxoplasmosis is especially serious for people who have weakened immune systems. For these people, they're at risk of developing:

- Brain inflammation, causing headaches, seizures, confusion and coma.
- A lung infection, causing cough, fever, and shortness of breath.
- An eye infection, causing blurry vision and eye pain.

When a fetus is infected, the symptoms may be mild or quite serious. Toxoplasmosis in an unborn baby can be life-threatening for the baby soon after birth. Most newborns with congenital toxoplasmosis may appear normal at birth but can develop signs and symptoms as they age. It's particularly important to check for involvement in their brain and eyes.

What are the Causes of Toxoplasmosis

T. gondii is theparasite that causes toxoplasmosis. You can catch it from contaminated meat that's raw or not thoroughly cooked. You can also get toxoplasmosis by drinking contaminated water. In rare cases, toxoplasmosis may be transmitted through a blood transfusion or a transplanted organ.

The parasite can also exist in feces. This means it can be found on some unwashed produce that has been contaminated with manure. Wash your produce thoroughly to prevent toxoplasmosis.

In the United States, the parasite is found in cat feces. Although T. gondii is found in nearly all warm-blooded animals, cats are the only known hosts. This means that the parasite's eggs only reproduce sexually in cats. The eggs exit the feline's body through excretion. Cats don't usually show symptoms of toxoplasmosis even though they're hosts.

People become infected with toxoplasmosis only if they ingest the parasite. This could happen when being exposed to contaminated cat feces. This is most likely when cleaning out a litter box without washing your hands afterward.

Pregnant women have an increased risk of passing toxoplasmosis to their unborn child in this manner. One to five days after it's shed.

It's very rare for humans to get toxoplasmosis from cats. Generally speaking, house cats that aren't allowed outside don't carry T. gondii. Wild cats or cats that live outside and hunt are more likely to be hosts of T. gondii.

In the United States, the most common way to get infected with the toxoplasmosis parasite is by eating raw meat or unwashed fruits and vegetables.

References

- "Infectious disease Campylobacter clinical Foodborne illnesses | CDC". Www.cdc.gov. Retrieved 2016-02-14
- Ryan, Kenneth James; Ray, C. George, eds. (2004). Sherris Medical Microbiology: An Introduction to Infectious Diseases (4th ed.). Mcgraw Hill. Pp. 378–80. ISBN 978-0-8385-8529-0
- "CAMPYLOBACTER : Food Safety Watch". Www.foodsafetywatch.org. Archived from the original on 2017-03-18. Retrieved 2017-03-17
- "Disease Listing, Staphylococcal Food Poisoning, General Info CDC Bacterial, Mycotic Diseases". Centers for Disease Control and Prevention. 29 March 2006. Retrieved 15 May 2012
- Giardiasis, health: healthline.com, Retrieved 25 February, 2019
- Baker-Austin, Craig; Oliver, James D. (February 2018). "Vibrio vulnificus: new insights into a deadly opportunistic pathogen". Environmental Microbiology. 20 (2): 423–430. Doi:10.1111/1462-2920.13955. ISSN 1462-2920. PMID 29027375
- Norovirus symptoms, health: healthline.com, Retrieved 16 January, 2019

6

Effects of Food Contamination

The presence of harmful microorganisms and chemicals in food that can cause food-borne illness is defined as food contamination. It can cause many diseases such as diarrhea, gastro-enteritis, nausea, etc. All these effects of food contamination have been carefully analyzed in this chapter.

Food Contamination

Food contamination is a commonly used term. However, only a few people are aware of the exact reasons for food contamination and its effects on your health. When food items are not handled or cooked safely, the disease-causing organisms such as bacteria, parasites, and viruses result in food contamination. The disease-causing parasites produce toxins that may also lead to food intoxication. In addition, the presence of pesticides, certain cleaning compounds, contaminate the food. The common reasons for food contamination are:

- Improper storing, handling and preparing food.
- Unhygienic hands and fingernails.
- Poor personal hygiene habits.
- Improperly cleaned or sanitized utensils.
- Contamination by flies, cockroaches, insects, and pests.

Different Types of Food Contamination

There are a number of reasons that can lead to food contamination. However, food contamination falls under four different categories which are:

- Biological contamination,
- Chemical contamination,
- Physical contamination,
- Cross-contamination.

Biological Contamination

Biological contamination is one of the common causes of food poisoning as well as spoilage. Contamination of food items by other living organisms is known as biological food contamination. During biological contamination, the harmful bacteria spread on foods that you consume. Even a single bacterium can multiply very quickly when they find ideal growth conditions. Not just bacteria, but also their process of multiplying can be quite harmful to humans. The common places where you can find bacteria are:

- Dust,
- Raw meat,
- The air,
- The human body,
- Pets and pests,
- Clothes of food handler,
- Kitchen clothes.

Physical Contamination

When harmful objects contaminate the food it leads to physical contamination. At times, food items can have both physical and biological contamination. Physical contaminants such as rats, hair, pests, glass or metals which can contaminate food and make it unhealthy. Some of the safety tips that you can follow when handling food items to prevent food contamination are:

- Hair-Tie your hair when handling food.
- Glass or Metal-Clean away cracked or broken crockery and utensils to avoid contamination.

- Fingernails-Keep your fingernails short or wear clean gloves when handling food.
- Dirt-Wash fruits and vegetables to remove dirt.
- Jewelry-Wear minimum jewelry when preparing food.

Chemical Contamination

Chemical contaminants are one of the serious sources of food contamination. These contaminants can also lead to food poisoning. Pesticides present in fruits and vegetables are one of the main sources of contamination. In addition, kitchen cleaning agents, food containers made of non-safe plastic, pest control products also lead to food contamination. Though we make it a point to wash fruits and vegetables thoroughly, however, plain water can't remove all the contaminants.

Cross-contamination

Many of us are not aware of cross contamination; however, this type of contamination can lead to a number of health problems. Cross-contamination takes place when pathogens are transported from any object that you use in the kitchen. Dirty kitchen clothes, unclean utensils, pests, raw food storage can lead to cross-contamination. Here are some of the ways to avoid cross-contamination:

- Personal Hygiene: Thoroughly wash your hands and face when handling food. Coughing, sneezing or even touching your hair can lead to cross contamination.
- Utensils: Use separate utensils to prepare different types of foods. Avoid using the same chopping board and knife for ready to eat foods.
- Storing Food: Make sure raw foods don't come in contact with ready to eat foods. Cover and store raw foods below cooked foods to prevent cross-contamination.
- Disposing Waste: Make sure you store and seal garbage correctly to prevent cross-contamination. Clean and sanitize the waste bins to prevent infestation risk.

Food Borne Illness

A disease caused by consuming contaminated food or drink is called foodborne disease. Myriad microbes and toxic substances can contaminate foods. There are more then 250 known foodborne diseases. The majority are infectious and are caused by bacteria, viruses, and parasites. Other foodborne diseases are essentially poisonings caused by toxins, chemicals contaminating the food. All foodborne microbes and toxins enter the body through the gastrointestinal tract and often causes the first symptoms there. Nausea, vomiting, abdominal cramps and diarrhea are frequent in foodborne diseases.

Many microbes can spread in more than one way, so it may not be immediately evident that a disease is foodborne. The distinction matters, because public health authorities need to know how a particular disease is spreading to take the appropriate steps to stop it. For example, infections with Escherichia coli O157:H7 (E. coli O157:H7) can be acquired through contaminated food, contaminated drinking water, contaminated swimming water, and from toddler to toddler at a day care center. Depending on which means of spread cause a case, the measures to stop other cases from occurring could range from removing contaminated food from stores, chlorinating a swimming pool, or closing a child day care center.

Diarrhea

Diarrhea, also spelled diarrhoea, is the condition of having at least three loose, liquid, or watery bowel movements each day. It often lasts for a few days and can result in dehydration due to fluid loss. Signs of dehydration often begin with loss of the normal stretchiness of the skin and irritable behaviour. This can progress to decreased urination, loss of skin color, a fast heart rate, and a decrease in responsiveness as it becomes more severe. Loose but non-watery stools in babies who are exclusively breastfed, however, are normal.

The most common cause is an infection of the intestines due to either a virus, bacteria, or parasite—a condition also known as gastroenteritis. These infections are often acquired from food or water that has been contaminated by feces, or directly from another person who is infected. The three types of diarrhea are: short duration watery diarrhea, short duration bloody diarrhea, and persistent diarrhea (lasting more than two weeks, which can be either watery or bloody). The short duration watery diarrhea may be due to cholera, although this is rare in the developed world. If blood is present, it is also known as dysentery. A number of non-infectious causes can result in diarrhea. These include lactose intolerance, irritable bowel syndrome, non-celiac gluten sensitivity, celiac disease, inflammatory bowel disease such as ulcerative colitis, hyperthyroidism, bile acid diarrhea, and a number of medications. In most cases, stool cultures to confirm the exact cause are not required.

Diarrhea can be prevented by improved sanitation, clean drinking water, and hand washing with soap. Breastfeeding for at least six months and vaccination against rotavirus is also recommended. Oral rehydration solution (ORS)—clean water with modest amounts of salts and sugar—is the treatment of choice. Zinc tablets are also recommended. These treatments have been estimated to have saved 50 million children in the past 25 years. When people have diarrhea it is recommended that they continue to eat healthy food and babies continue to be breastfed. If commercial ORS are not available, homemade solutions may be used. In those with severe dehydration, intravenous fluids may be required. Most cases; however, can be managed well with fluids by mouth. Antibiotics, while rarely used, may be recommended in a few cases such as those who have bloody diarrhea and a high fever, those with severe diarrhea following travelling, and those who grow specific bacteria or parasites in their stool. Loperamide may help decrease the number of bowel movements but is not recommended in those with severe disease.

About 1.7 to 5 billion cases of diarrhea occur per year. It is most common in developing countries, where young children get diarrhea on average three times a year. Total deaths from diarrhea are estimated at 1.26 million in 2013—down from 2.58 million in 1990. In 2012, it was the second most common cause of deaths in children younger than five (0.76 million or 11%). Frequent episodes of diarrhea are also a common cause of malnutrition and the most common cause in those younger than five years of age. Other long term problems that can result include stunted growth and poor intellectual development.

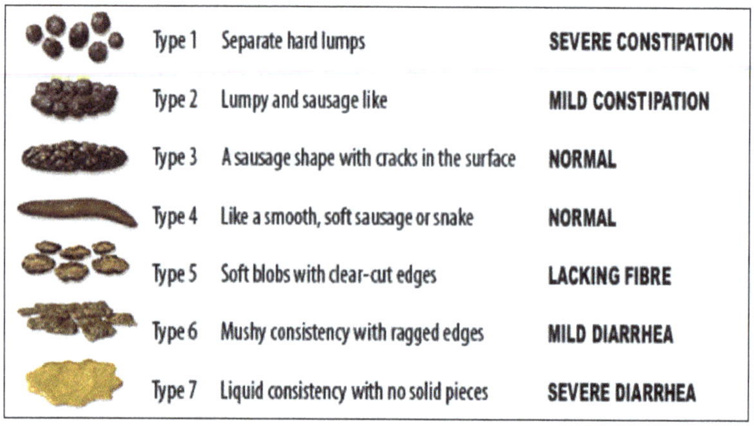

Bristol stool chart.

Diarrhea is defined by the World Health Organization as having three or more loose or liquid stools per day, or as having more stools than is normal for that person.

Acute diarrhea is defined as an abnormally frequent discharge of semisolid or fluid fecal matter from the bowel, lasting less than 14 days, by World Gastroenterology Organization.

Secretory Diarrhea

Secretory diarrhea means that there is an increase in the active secretion, or there is an inhibition of absorption. There is little to no structural damage. The most common cause of this type of diarrhea is a cholera toxin that stimulates the secretion of anions, especially chloride ions (Cl^-). Therefore, to maintain a charge balance in the gastrointestinal tract, sodium (Na^+) is carried with it, along with water. In this type of diarrhea intestinal fluid secretion is isotonic with plasma even during fasting. It continues even when there is no oral food intake.

Osmotic Diarrhea

Osmotic diarrhea occurs when too much water is drawn into the bowels. If a person drinks solutions with excessive sugar or excessive salt, these can draw water from the body into the bowel and cause osmotic diarrhea. Osmotic diarrhea can also result from maldigestion, e.g. pancreatic disease or coeliac disease in which the nutrients are left in the lumen to pull in water. Or it can be caused by osmotic laxatives (which work to alleviate constipation by drawing water into the bowels). In healthy individuals, too much magnesium or vitamin C or undigested lactose can produce osmotic diarrhea and distention of the bowel. A person who has lactose intolerance can have difficulty absorbing lactose after an extraordinarily high intake of dairy products. In persons who have fructose malabsorption, excess fructose intake can also cause diarrhea. High-fructose foods that also have a high glucose content are more absorbable and less likely to cause diarrhea. Sugar alcohols such as sorbitol (often found in sugar-free foods) are difficult for the body to absorb and, in large amounts, may lead to osmotic diarrhea. In most of these cases, osmotic diarrhea stops when the offending agent, e.g. milk or sorbitol, is stopped.

Exudative Diarrhea

Exudative diarrhea occurs with the presence of blood and pus in the stool. This occurs with inflammatory bowel diseases, such as Crohn's disease or ulcerative colitis, and other severe infections such as *E. coli* or other forms of food poisoning.

Inflammatory Diarrhea

Inflammatory diarrhea occurs when there is damage to the mucosal lining or brush border, which leads to a passive loss of protein-rich fluids and a decreased ability to absorb these lost fluids. Features of all three of the other types of diarrhea can be found in this type of diarrhea. It can be caused by bacterial infections, viral infections, parasitic infections, or autoimmune problems such as inflammatory bowel diseases. It can also be caused by tuberculosis, colon cancer, and enteritis.

Dysentery

If there is blood visible in the stools, it is also known as dysentery. The blood is a trace of an invasion of bowel tissue. Dysentery is a symptom of, among others, Shigella, Entamoeba histolytica, and Salmonella.

Health Effects

Diarrheal disease may have a negative impact on both physical fitness and mental development. "Early childhood malnutrition resulting from any cause reduces physical fitness and work productivity in adults," and diarrhea is a primary cause of childhood malnutrition. Further, evidence suggests that diarrheal disease has significant impacts on mental development and health; it has been shown that, even when controlling for helminth infection and early breastfeeding, children who had experienced severe diarrhea had significantly lower scores on a series of tests of intelligence.

Diarrhea can cause electrolyte imbalances, kidney impairment, dehydration, and defective immune system responses. When oral drugs are administered, the efficiency of the drug is to produce a therapeutic effect and the lack of this effect may be due to the medication travelling too quickly through the digestive system, limiting the time that it can be absorbed. Clinicians try to treat the diarrheas by reducing the dosage of medication, changing the dosing schedule, discontinuation of the drug, and rehydration. The interventions to control the diarrhea are not often effective. Diarrhea can have a profound effect on the quality of life because fecal incontinence is one of the leading factors for placing older adults in long term care facilities (nursing homes).

Causes

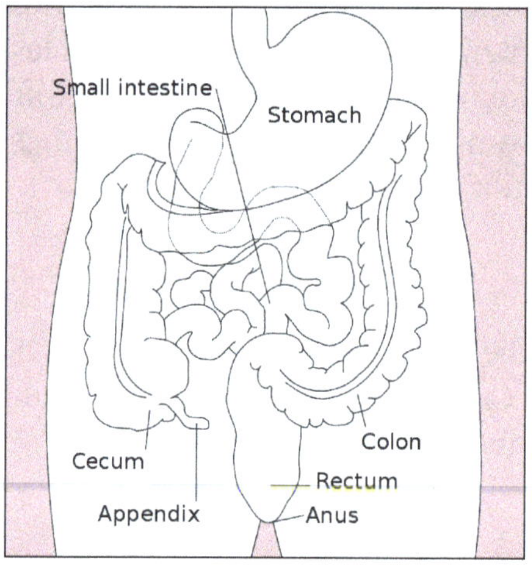

Diagram of the human gastrointestinal tract.

Acute diarrhea is most commonly due to viral gastroenteritis with rotavirus, which accounts for 40% of cases in children under five. In travelers, however, bacterial infections predominate. Various toxins such as mushroom poisoning and drugs can also cause acute diarrhea.

Chronic diarrhea can be the part of the presentations of a number of chronic medical conditions affecting the intestine. Common causes include ulcerative colitis, Crohn disease, microscopic colitis, celiac disease, irritable bowel syndrome, and bile acid malabsorption.

Infections

There are many causes of infectious diarrhea, which include viruses, bacteria and parasites. Infectious diarrhea is frequently referred to as gastroenteritis. Norovirus is the most common cause of viral diarrhea in adults, but rotavirus is the most common cause in children under five years old. Adenovirus types 40 and 41, and astroviruses cause a significant number of infections. Shiga-toxin producing Escherichia coli, such as E coli o157:h7, are the most common cause of infectious bloody diarrhea in the United States.

Campylobacter spp. are a common cause of bacterial diarrhea, but infections by Salmonella spp., Shigella spp. and some strains of Escherichia coli are also a frequent cause.

In the elderly, particularly those who have been treated with antibiotics for unrelated infections, a toxin produced by Clostridioides difficile often causes severe diarrhea.

Parasites, particularly protozoa e.g., Cryptosporidium spp., Giardia spp., Entamoeba histolytica, Blastocystis spp., Cyclospora cayetanensis, are frequently the cause of diarrhea that involves chronic infection. The broad-spectrum antiparasitic agent nitazoxanide has shown efficacy against many diarrhea-causing parasites.

Other infectious agents, such as parasites or bacterial toxins, may exacerbate symptoms. In sanitary living conditions where there is ample food and a supply of clean water, an otherwise healthy person usually recovers from viral infections in a few days. However, for ill or malnourished individuals, diarrhea can lead to severe dehydration and can become life-threatening.

Sanitation

Poverty often leads to unhygienic living conditions, as in this community in the Indian Himalayas. Such conditions promote contraction of diarrheal diseases, as a result of poor sanitation and hygiene.

Open defecation is a leading cause of infectious diarrhea leading to death. Poverty is a good indicator of the rate of infectious diarrhea in a population. This association does not stem from poverty itself, but rather from the conditions under which impoverished people live. The absence of certain resources compromises the ability of the poor to defend themselves against infectious diarrhea. "Poverty is associated with poor housing, crowding, dirt floors, lack of access to clean water or to sanitary disposal of fecal waste (sanitation), cohabitation with domestic animals that may carry human pathogens, and a lack of refrigerated storage for food, all of which increase the frequency of diarrhea. Poverty also restricts the ability to provide age-appropriate, nutritionally balanced diets or to modify diets when diarrhea develops so as to mitigate and repair nutrient losses. The impact is exacerbated by the lack of adequate, available, and affordable medical care."

One of the most common causes of infectious diarrhea is a lack of clean water. Often, improper fecal disposal leads to contamination of groundwater. This can lead to widespread infection among a population, especially in the absence of water filtration or purification. Human feces contains a variety of potentially harmful human pathogens.

Nutrition

Proper nutrition is important for health and functioning, including the prevention of infectious diarrhea. It is especially important to young children who do not have a fully developed immune system. Zinc deficiency, a condition often found in children in developing countries can, even in mild cases, have a significant impact on the development and proper functioning of the human immune system. Indeed, this relationship between zinc deficiency and reduced immune functioning corresponds with an increased severity of infectious diarrhea. Children who have lowered levels of zinc have a greater number of instances of diarrhea, severe diarrhea, and diarrhea associated with fever. Similarly, vitamin A deficiency can cause an increase in the severity of diarrheal episodes. However, there is some discrepancy when it comes to the impact of vitamin A deficiency on the rate of disease. While some argue that a relationship does not exist between the rate of disease and vitamin A status, Others suggest an increase in the rate associated with deficiency. Given that estimates suggest 127 million preschool children worldwide are vitamin A deficient, this population has the potential for increased risk of disease contraction.

Malabsorption

Malabsorption is the inability to absorb food fully, mostly from disorders in the small bowel, but also due to maldigestion from diseases of the pancreas.

Causes include:

- Enzyme deficiencies or mucosal abnormality, as in food allergy and food intolerance, e.g. celiac disease (gluten intolerance), lactose intolerance (intolerance to milk sugar, common in non-Europeans), and fructose malabsorption,
- Pernicious anemia, or impaired bowel function due to the inability to absorb vitamin B12,
- Loss of pancreatic secretions, which may be due to cystic fibrosis or pancreatitis,
- Structural defects, like short bowel syndrome (surgically removed bowel) and radiation fibrosis, such as usually follows cancer treatment and other drugs, including agents used in chemotherapy,
- Certain drugs, like orlistat, which inhibits the absorption of fat.

Inflammatory Bowel Disease

The two overlapping types here are of unknown origin:

- Ulcerative colitis is marked by chronic bloody diarrhea and inflammation mostly affects the distal colon near the rectum.
- Crohn's disease typically affects fairly well demarcated segments of bowel in the colon and often affects the end of the small bowel.

Irritable Bowel Syndrome

Another possible cause of diarrhea is irritable bowel syndrome (IBS), which usually presents with abdominal discomfort relieved by defecation and unusual stool (diarrhea or constipation) for at least three days a week over the previous three months. Symptoms of diarrhea-predominant IBS can be managed through a combination of dietary changes, soluble fiber supplements and medications such as loperamide or codeine. About 30% of patients with diarrhea-predominant IBS have bile acid malabsorption diagnosed with an abnormal SeHCAT test.

Other Diseases

Diarrhea can be caused by other diseases and conditions, namely:

- Chronic ethanol ingestion,
- Hyperthyroidism,
- Certain medications,

- Bile acid malabsorption,
- Ischemic bowel disease: This usually affects older people and can be due to blocked arteries,
- Microscopic colitis, a type of inflammatory bowel disease where changes are seen only on histological examination of colonic biopsies,
- Bile salt malabsorption (primary bile acid diarrhea) where excessive bile acids in the colon produce a secretory diarrhea,
- Hormone-secreting tumors: some hormones, e.g. serotonin), can cause diarrhea if excreted in excess (usually from a tumor),
- Chronic mild diarrhea in infants and toddlers may occur with no obvious cause and with no other ill effects; this condition is called toddler's diarrhea,
- Environmental enteropathy,
- Radiation enteropathy following treatment for pelvic and abdominal cancers.

Medications

Some medications, such as the penicillum can cause diarrhea. Over 700 medications are known to cause diarrhea. The classes of medications that are known to cause diarrhea are laxatives, antacids, heartburn medications, antibiotics, anti-neoplastic drugs, anti-inflammatories as well as many dietary supplements.

Pathophysiology

Evolution

Ion transporters targeted by enteric infections	
Function	Transporter
Absorption	NHE, SGLT1, ENaC, DRA
Secretion	CaCC, NKCC1, CFTR
Absorption and secretion	Sodium potassium ATPase

According to two researchers, Nesse and Williams, diarrhea may function as an evolved expulsion defense mechanism. As a result, if it is stopped, there might be a delay in recovery. They cite in support of this argument research published in 1973 that found that treating *Shigella* with the anti-diarrhea drug (Co-phenotrope, Lomotil) caused people to stay feverish twice as long as those not so treated. The researchers indeed themselves observed that: "Lomotil may be contraindicated in shigellosis. Diarrhea may represent a defense mechanism".

Diagnostic Approach

The following types of diarrhea may indicate further investigation is needed:

- In infants,

- Moderate or severe diarrhea in young children,
- Associated with blood,
- Continues for more than two days,
- Associated non-cramping abdominal pain, fever, weight loss, etc.,
- In travelers,
- In food handlers, because of the potential to infect others,
- In institutions such as hospitals, child care centers, or geriatric and convalescent homes.

A severity score is used to aid diagnosis in children.

Chronic Diarrhea

When diarrhea lasts for more than four weeks a number of further tests may be recommended including:

- Complete blood count and a ferritin if anemia is present,
- Thyroid stimulating hormone,
- Tissue transglutaminase for celiac disease,
- Fecal calprotectin to exclude inflammatory bowel disease,
- Stool tests for ova and parasites as well as for *Clostridioides difficile,*
- A colonoscopy or fecal immunochemical testing for cancer, including biopsies to detect microscopic colitis,
- Testing for bile acid diarrhea with SeHCAT, 7α-hydroxy-4-cholesten-3-one or fecal bile acids depending on availability,
- Hydrogen breath test looking for lactose intolerance,
- Further tests if immunodeficiency, pelvic radiation disease or small intestinal bacterial overgrowth suspected.

Prevention

Sanitation

Numerous studies have shown that improvements in drinking water and sanitation (WASH) lead to decreased risks of diarrhoea. Such improvements might include for example use of water filters, provision of high-quality piped water and sewer connections.

In institutions, communities, and households, interventions that promote hand washing with soap lead to significant reductions in the incidence of diarrhea. The same applies to preventing open defecation at a community-wide level and providing access to improved sanitation. This includes use of toilets and implementation of the entire sanitation chain connected to the toilets (collection, transport, disposal or reuse of human excreta).

Hand Washing

Basic sanitation techniques can have a profound effect on the transmission of diarrheal disease. The implementation of hand washing using soap and water, for example, has been experimentally shown to reduce the incidence of disease by approximately 42–48%. Hand washing in developing countries, however, is compromised by poverty as acknowledged by the CDC: "Handwashing is integral to disease prevention in all parts of the world; however, access to soap and water is limited in a number of less developed countries. This lack of access is one of many challenges to proper hygiene in less developed countries." Solutions to this barrier require the implementation of educational programs that encourage sanitary behaviours.

Water

Given that water contamination is a major means of transmitting diarrheal disease, efforts to provide clean water supply and improved sanitation have the potential to dramatically cut the rate of disease incidence. In fact, it has been proposed that we might expect an 88% reduction in child mortality resulting from diarrheal disease as a result of improved water sanitation and hygiene. Similarly, a meta-analysis of numerous studies on improving water supply and sanitation shows a 22–27% reduction in disease incidence, and a 21–30% reduction in mortality rate associated with diarrheal disease.

Chlorine treatment of water, for example, has been shown to reduce both the risk of diarrheal disease, and of contamination of stored water with diarrheal pathogens.

Vaccination

Immunization against the pathogens that cause diarrheal disease is a viable prevention strategy, however it does require targeting certain pathogens for vaccination. In the case of Rotavirus, which was responsible for around 6% of diarrheal episodes and 20% of diarrheal disease deaths in the children of developing countries, use of a Rotavirus vaccine in trials in 1985 yielded a slight (2–3%) decrease in total diarrheal disease incidence, while reducing overall mortality by 6–10%. Similarly, a Cholera vaccine showed a strong reduction in morbidity and mortality, though the overall impact of vaccination was minimal as Cholera is not one of the major causative pathogens of diarrheal disease. Since this time, more effective vaccines have been developed that have the potential to save many thousands of lives in developing nations, while reducing the overall cost of treatment, and the costs to society.

A rotavirus vaccine decrease the rates of diarrhea in a population. New vaccines against rotavirus, Shigella, Enterotoxigenic Escherichia coli (ETEC), and cholera are under development, as well as other causes of infectious diarrhea.

Nutrition

Dietary deficiencies in developing countries can be combated by promoting better eating practices. Zinc supplementation proved successful showing a significant decrease in the incidence of diarrheal disease compared to a control group. The majority of the literature suggests that vitamin A supplementation is advantageous in reducing disease incidence. Development of a supplementation strategy should take into consideration the fact that vitamin A supplementation was less effective in reducing diarrhea incidence when compared to vitamin A and zinc supplementation, and that the latter strategy was estimated to be significantly more cost effective.

Breastfeeding

Breastfeeding practices have been shown to have a dramatic effect on the incidence of diarrheal disease in poor populations. Studies across a number of developing nations have shown that those who receive exclusive breastfeeding during their first 6 months of life are better protected against infection with diarrheal diseases. One study in Brazil found that non-breastfed infants were 14 times more likely to die from diarrhea than exclusively breastfed infants. Exclusive breastfeeding is currently recommended for the first six months of an infant's life by the WHO, with continued breastfeeding until at least two years of age.

Probiotics decrease the risk of diarrhea in those taking antibiotics.

Management

In many cases of diarrhea, replacing lost fluid and salts is the only treatment needed. This is usually by mouth – oral rehydration therapy – or, in severe cases, intravenously. Diet restrictions such as the BRAT diet are no longer recommended. Research does not support the limiting of milk to children as doing so has no effect on duration of diarrhea. To the contrary, WHO recommends that children with diarrhea continue to eat as sufficient nutrients are usually still absorbed to support continued growth and weight gain, and that continuing to eat also speeds up recovery of normal intestinal functioning. CDC recommends that children and adults with cholera also continue to eat.

Medications such as loperamide (Imodium) and bismuth subsalicylate may be beneficial; however they may be contraindicated in certain situations.

Fluids

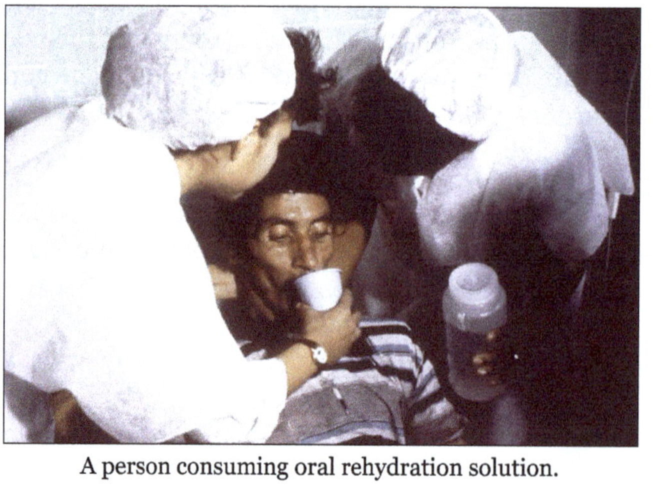

A person consuming oral rehydration solution.

Oral rehydration solution (ORS) (a slightly sweetened and salty water) can be used to prevent dehydration. Standard home solutions such as salted rice water, salted yogurt drinks, vegetable and chicken soups with salt can be given. Home solutions such as water in which cereal has been cooked, unsalted soup, green coconut water, weak tea (unsweetened), and unsweetened fresh fruit juices can have from half a teaspoon to full teaspoon of salt (from one-and-a-half to three grams) added per liter. Clean plain water can also be one of several fluids given. There are commercial solutions such as Pedialyte, and relief agencies such as UNICEF widely distribute packets of salts and sugar. A WHO publication for physicians recommends a homemade ORS consisting of one liter water with one teaspoon salt (3 grams) and two tablespoons sugar (18 grams) added (approximately the "taste of tears"). Rehydration Project recommends adding the same amount of sugar but only one-half a teaspoon of salt, stating that this more dilute approach is less risky with very little loss of effectiveness. Both agree that drinks with too much sugar or salt can make dehydration worse.

Appropriate amounts of supplemental zinc and potassium should be added if available. But the availability of these should not delay rehydration. As WHO points out, the most important thing is to begin preventing dehydration as early as possible. In another example of prompt ORS hopefully preventing dehydration, CDC recommends for the treatment of cholera continuing to give Oral Rehydration Solution during travel to medical treatment.

Vomiting often occurs during the first hour or two of treatment with ORS, especially if a child drinks the solution too quickly, but this seldom prevents successful rehydration since most of the fluid is still absorbed. WHO recommends that if a child vomits, to wait five or ten minutes and then start to give the solution again more slowly.

Drinks especially high in simple sugars, such as soft drinks and fruit juices, are not recommended in children under five as they may *increase* dehydration. A too rich solution in the gut draws water from the rest of the body, just as if the person were to drink sea water. Plain water may be used if more specific and effective ORT preparations are unavailable or are not palatable. Additionally, a mix of both plain water and drinks perhaps too rich in sugar and salt can alternatively be given to the same person, with the goal of providing a medium amount of sodium overall. A nasogastric tube can be used in young children to administer fluids if warranted.

Eating

The WHO recommends a child with diarrhea continue to be fed. Continued feeding speeds the recovery of normal intestinal function. In contrast, children whose food is restricted have diarrhea of longer duration and recover intestinal function more slowly. The WHO states "Food should *never* be withheld and the child's usual foods should *not* be diluted. Breastfeeding should *always* be continued." And in the specific example of cholera, CDC also makes the same recommendation. Breastfed infants with diarrhea often choose to breastfeed more, and should be encouraged to do so. In young children who are not breast-fed and live in the developed world, a lactose-free diet may be useful to speed recovery.

Medications

While antibiotics are beneficial in certain types of acute diarrhea, they are usually not used except in specific situations. There are concerns that antibiotics may increase the risk of hemolytic uremic syndrome in people infected with Escherichia coli O157:H7. In resource-poor countries, treatment with antibiotics may be beneficial. However, some bacteria are developing antibiotic resistance, particularly Shigella. Antibiotics can also cause diarrhea, and antibiotic-associated diarrhea is the most common adverse effect of treatment with general antibiotics.

While bismuth compounds (Pepto-Bismol) decreased the number of bowel movements in those with travelers' diarrhea, they do not decrease the length of illness. Anti-motility agents like loperamide are also effective at reducing the number of stools but not the duration of disease. These agents should be used only if bloody diarrhea is not present.

Diosmectite, a natural aluminomagnesium silicate clay, is effective in alleviating symptoms of acute diarrhea in children, and also has some effects in chronic functional diarrhea, radiation-induced diarrhea, and chemotherapy-induced diarrhea.

Bile acid sequestrants such as cholestyramine can be effective in chronic diarrhea due to bile acid malabsorption. Therapeutic trials of these drugs are indicated in chronic diarrhea if bile acid malabsorption cannot be diagnosed with a specific test, such as SeHCAT retention.

Alternative Therapies

Zinc supplementation may benefit children over six months old with diarrhea in areas with high rates of malnourishment or zinc deficiency. This supports the World Health Organization guidelines for zinc, but not in the very young.

Probiotics reduce the duration of symptoms by one day and reduced the chances of symptoms lasting longer than four days by 60%. The probiotic lactobacillus can help prevent antibiotic-associated diarrhea in adults but possibly not children. For those with lactose intolerance, taking digestive enzymes containing lactase when consuming dairy products often improves symptoms.

Epidemiology

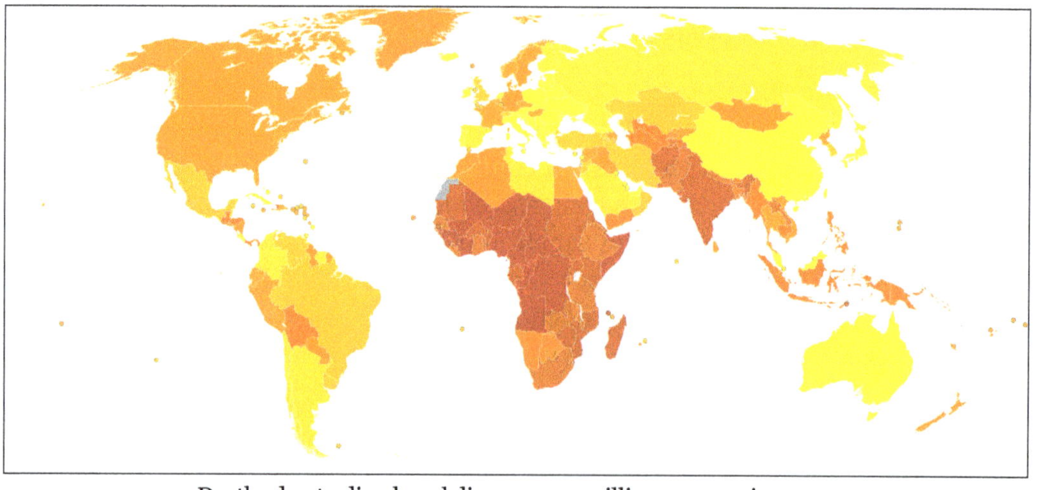

Deaths due to diarrhoeal diseases per million persons in 2012.

Worldwide in 2004, approximately 2.5 billion cases of diarrhea occurred, which resulted in 1.5 million deaths among children under the age of five. Greater than half of these were in Africa and South Asia. This is down from a death rate of 4.5 million in 1980 for gastroenteritis. Diarrhea remains the second leading cause of infant mortality (16%) after pneumonia (17%) in this age group.

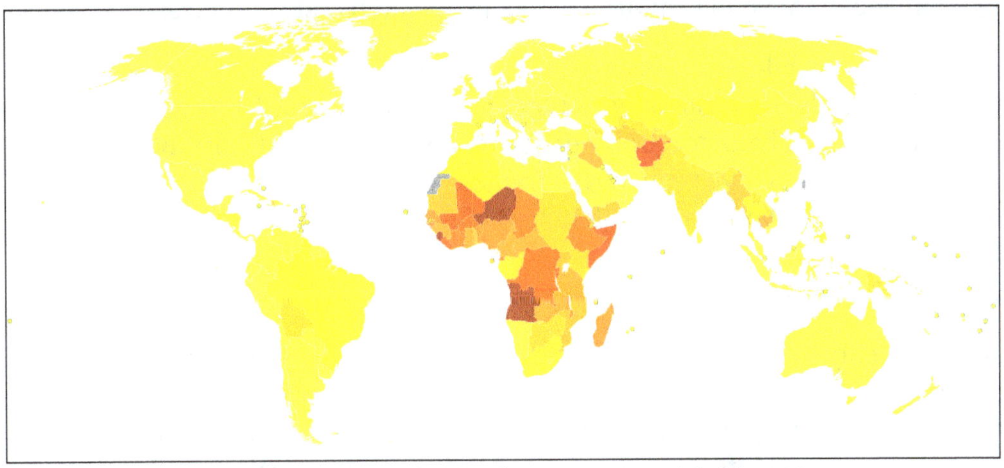

Disability-adjusted life year for diarrhea per 100,000 inhabitants in 2004.

no data; < 500; 500–1000; 1000–1500; 1500–2000; 2000–2500; 2500–3000; 3000–3500; 3500–4000; 4000–4500; 4500–5000; 5000–6000; > 6000

The majority of such cases occur in the developing world, with over half of the recorded cases of childhood diarrhea occurring in Africa and Asia, with 696 million and 1.2 billion cases, respectively, compared to only 480 million in the rest of the world.

Infectious diarrhea resulted in about 0.7 million deaths in children under five years old in 2011 and 250 million lost school days. In the Americas, diarrheal disease accounts for a total of 10% of deaths among children aged 1–59 months while in South East Asia, it accounts for 31.3% of deaths. It is estimated that around 21% of child mortalities in developing countries are due to diarrheal disease.

Gastroenteritis

Gastroenteritis, also known as infectious diarrhea, is inflammation of the gastrointestinal tract—the stomach and small intestine. Symptoms may include diarrhea, vomiting and abdominal pain. Fever, lack of energy and dehydration may also occur. This typically lasts less than two weeks. It is not related to influenza, though it has erroneously been called the "stomach flu".

Gastroenteritis is usually caused by viruses. However, bacteria, parasites, and fungus can also cause gastroenteritis. In children, rotavirus is the most common cause of severe disease. In adults, norovirus and *Campylobacter* are common causes. Eating improperly prepared food, drinking contaminated water or close contact with a person who is infected can spread the disease. Treatment is generally the same with or without a definitive diagnosis, so testing to confirm is usually not needed.

Prevention includes hand washing with soap, drinking clean water, proper disposal of human waste and breastfeeding babies instead of using formula. The rotavirus vaccine is recommended as a prevention for children. Treatment involves getting enough fluids. For mild or moderate cases, this can typically be achieved by drinking oral rehydration solution (a combination of water, salts and sugar). In those who are breastfed, continued breastfeeding is recommended. For more severe cases, intravenous fluids may be needed. Fluids may also be given by a nasogastric tube. Zinc supplementation is recommended in children. Antibiotics are generally not needed. However, antibiotics are recommended for young children with a fever and bloody diarrhea.

In 2015, there were two billion cases of gastroenteritis, resulting in 1.3 million deaths globally. Children and those in the developing world are affected the most. In 2011, there were about 1.7 billion cases, resulting in about 700,000 deaths of children under the age of five. In the developing world, children less than two years of age frequently get six or more infections a year. It is less common in adults, partly due to the development of immunity.

Signs and Symptoms

Gastroenteritis usually involves both diarrhea and vomiting. Sometimes, only one or the other is present. This may be accompanied by abdominal cramps. Signs and symptoms usually begin 12–72 hours after contracting the infectious agent. If due to a virus, the condition usually resolves within one week. Some viral infections also involve fever, fatigue, headache and muscle pain. If the stool is bloody, the cause is less likely to be viral and more likely to be bacterial. Some bacterial infections cause severe abdominal pain and may persist for several weeks.

Children infected with rotavirus usually make a full recovery within three to eight days. However, in poor countries treatment for severe infections is often out of reach and persistent diarrhea is common. Dehydration is a common complication of diarrhea. Severe dehydration in children may be recognized if the skin color and position returns slowly when pressed. This is called "prolonged capillary refill" and "poor skin turgor". Abnormal breathing is another sign of severe dehydration. Repeat infections are typically seen in areas with poor sanitation, and malnutrition. Stunted growth and long-term cognitive delays can result.

Reactive arthritis occurs in 1% of people following infections with Campylobacter species. Guillain–Barré syndrome occurs in 0.1%. Hemolytic uremic syndrome (HUS) may occur due to infection with Shiga toxin-producing Escherichia coli or Shigella species. HUS causes low platelet counts, poor kidney function, and low red blood cell count (due to their breakdown). Children are more predisposed to getting HUS than adults. Some viral infections may produce benign infantile seizures.

Cause

Viruses (particularly rotavirus) and the bacteria Escherichia coli and Campylobacter species are the primary causes of gastroenteritis. There are, however, many other infectious agents that can cause this syndrome including parasites and fungus. Non-infectious causes are seen on occasion, but they are less likely than a viral or bacterial cause. Risk of infection is higher in children due to their lack of immunity. Children are also at higher risk because they are less likely to practice good hygiene habits. Children living in areas without easy access to water and soap are especially vulnerable.

Viral

Rotavirus, norovirus, adenovirus, and astrovirus are known to cause viral gastroenteritis. Rotavirus is the most common cause of gastroenteritis in children, and produces similar rates in both the developed and developing world. Viruses cause about 70% of episodes of infectious diarrhea in the pediatric age group. Rotavirus is a less common cause in adults due to acquired immunity. Norovirus is the cause in about 18% of all cases.

Norovirus is the leading cause of gastroenteritis among adults in America, causing greater than 90% of outbreaks. These localized epidemics typically occur when groups of people spend time in close physical proximity to each other, such as on cruise ships, in hospitals, or in restaurants. People may remain infectious even after their diarrhea has ended. Norovirus is the cause of about 10% of cases in children.

Bacterial

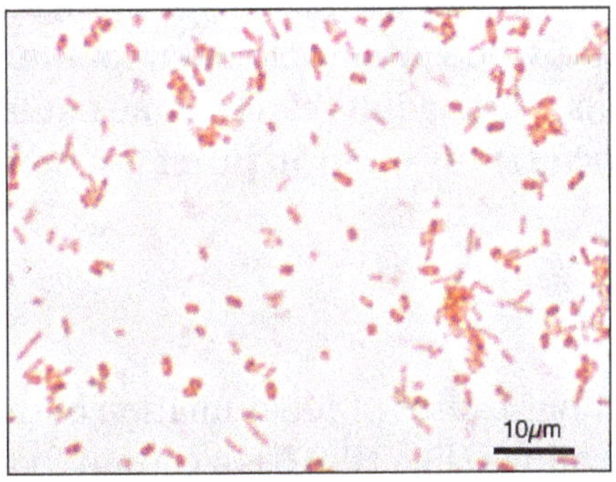

Salmonella enterica serovar Typhimurium (ATCC 14028) as seen with a microscope at 1000 fold magnification and following Gram staining.

In the developed world Campylobacter jejuni is the primary cause of bacterial gastroenteritis, with half of these cases associated with exposure to poultry. In children, bacteria are the cause in about 15% of cases, with the most common types being Escherichia coli, Salmonella, Shigella, and Campylobacter species. If food becomes contaminated with bacteria and remains at room temperature for a period of several hours, the bacteria multiply and increase the risk of infection in those who consume the food. Some foods commonly associated with illness include raw or undercooked meat, poultry, seafood, and eggs; raw sprouts; unpasteurized milk and soft cheeses; and fruit and vegetable juices. In the developing world, especially sub-Saharan Africa and Asia, cholera is a common cause of gastroenteritis. This infection is usually transmitted by contaminated water or food.

Toxigenic Clostridium difficile is an important cause of diarrhea that occurs more often in the elderly. Infants can carry these bacteria without developing symptoms. It is a common cause of diarrhea in those who are hospitalized and is frequently associated with antibiotic use. Staphylococcus aureus infectious diarrhea may also occur in those who have used antibiotics. Acute "traveler's diarrhea" is usually a type of bacterial gastroenteritis, while the persistent form is usually parasitic. Acid-suppressing medication appears to increase the risk of significant infection after exposure to a number of organisms, including Clostridium difficile, Salmonella, and Campylobacter species. The risk is greater in those taking proton pump inhibitors than with H2 antagonists.

Parasitic

A number of parasites can cause gastroenteritis. Giardia lamblia is most common, but Entamoeba histolytica, Cryptosporidium spp., and other species have also been implicated. As a group, these agents comprise about 10% of cases in children. Giardia occurs more commonly in the developing world, but this type of illness can occur nearly everywhere. It occurs more commonly in persons who have traveled to areas with high prevalence, children who attend day care, men who have sex with men, and following disasters.

Transmission

Transmission may occur from drinking contaminated water or when people share personal objects. Water quality typically worsens during the rainy season and outbreaks are more common at this time. In areas with four seasons, infections are more common in the winter. Worldwide, bottle-feeding of babies with improperly sanitized bottles is a significant cause. Transmission rates are also related to poor hygiene, (especially among children), in crowded households, and in those with poor nutritional status. Adults who have developed immunities may still carry certain organisms without exhibiting symptoms. Thus, adults can become natural reservoirs of certain diseases. While some agents (such as *Shigella*) only occur in primates, others (such as *Giardia*) may occur in a wide variety of animals.

Non-infectious

There are a number of non-infectious causes of inflammation of the gastrointestinal tract. Some of the more common include medications (like NSAIDs), certain foods such as lactose (in those who are intolerant), and gluten (in those with celiac disease). Crohn's disease is also a non-infectious source of (often severe) gastroenteritis. Disease secondary to toxins may also occur. Some food-related conditions associated with nausea, vomiting, and diarrhea include: Ciguatera poisoning due to consumption of contaminated predatory fish, scombroid associated with the consumption of certain types of spoiled fish, tetrodotoxin poisoning from the consumption of puffer fish among others, and botulism typically due to improperly preserved food.

In the United States, rates of emergency department use for noninfectious gastroenteritis dropped 30% from 2006 until 2011. Of the twenty most common conditions seen in the emergency department, rates of noninfectious gastroenteritis had the largest decrease in visits in that time period.

Pathophysiology

Gastroenteritis is defined as vomiting or diarrhea due to inflammation of the small or large bowel, often due to infection. The changes in the small bowel are typically noninflammatory, while the ones in the large bowel are inflammatory. The number of pathogens required to cause an infection varies from as few as one (for Cryptosporidium) to as many as 108 (for Vibrio cholerae).

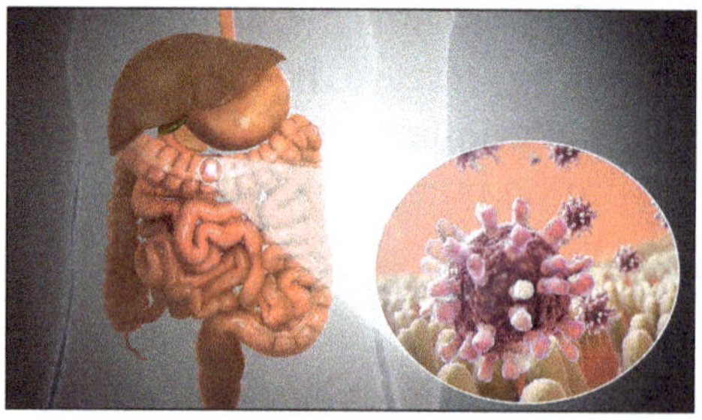

Gastroenteritis caused due to intestinal infection.

Prevention

Lifestyle

A supply of easily accessible uncontaminated water and good sanitation practices are important for reducing rates of infection and clinically significant gastroenteritis. Personal measures (such as hand washing with soap) have been found to decrease rates of gastroenteritis in both the developing and developed world by as much as 30%. Alcohol-based gels may also be effective. Food or drink that is thought to be contaminated should be avoided. Breastfeeding is important, especially in places with poor hygiene, as is improvement of hygiene generally. Breast milk reduces both the frequency of infections and their duration.

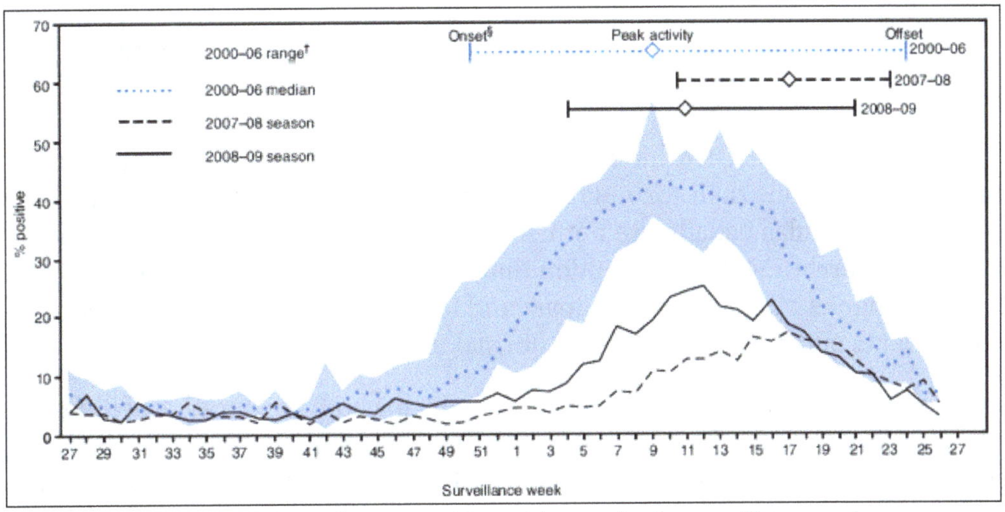

Percentage of rotavirus tests with positive results, by surveillance week.

Vaccination

Due to both its effectiveness and safety, in 2009 the World Health Organization recommended that the rotavirus vaccine be offered to all children globally. Two commercial rotavirus vaccines exist and several more are in development. In Africa and Asia these vaccines reduced severe disease among infants and countries that have put in place national immunization programs have seen a decline in the rates and severity of disease. This vaccine may also prevent illness in non-vaccinated children by reducing the number of circulating infections. Since 2000, the implementation of a rotavirus vaccination program in the United States has substantially decreased the number of cases of diarrhea by as much as 80 percent. The first dose of vaccine should be given to infants between 6 and 15 weeks of age. The oral cholera vaccine has been found to be 50–60% effective over 2 years.

Management

Gastroenteritis is usually an acute and self-limiting disease that does not require medication. The preferred treatment in those with mild to moderate dehydration is oral rehydration therapy (ORT). For children at risk of dehydration from vomiting, taking a single dose of the anti vomiting medication metoclopramide or ondansetron, may be helpful, and butylscopolamine is useful in treating abdominal pain.

Rehydration

The primary treatment of gastroenteritis in both children and adults is rehydration. This is preferably achieved by drinking rehydration solution, although intravenous delivery may be required if there is a decreased level of consciousness or if dehydration is severe. Drinking replacement therapy products made with complex carbohydrates (i.e. those made from wheat or rice) may be superior to those based on simple sugars. Drinks especially high in simple sugars, such as soft drinks and fruit juices, are not recommended in children under 5 years of age as they may *increase* diarrhea. Plain water may be used if more specific ORT preparations are unavailable or the person is not willing to drink them. A nasogastric tube can be used in young children to administer fluids if warranted. In those who require intravenous fluids, one to four hours' worth is often sufficient.

Dietary

It is recommended that breast-fed infants continue to be nursed in the usual fashion, and that formula-fed infants continue their formula immediately after rehydration with ORT. Lactose-free or lactose-reduced formulas usually are not necessary. Children should continue their usual diet during episodes of diarrhea with the exception that foods high in simple sugars should be avoided. The BRAT diet (bananas, rice, applesauce, toast and tea) is no longer recommended, as it contains insufficient nutrients and has no benefit over normal feeding.

Some probiotics have been shown to be beneficial in reducing both the duration of illness and the frequency of stools. They may also be useful in preventing and treating antibiotic associated diarrhea. Fermented milk products (such as yogurt) are similarly beneficial. Zinc supplementation appears to be effective in both treating and preventing diarrhea among children in the developing world.

Antiemetics

Antiemetic medications may be helpful for treating vomiting in children. Ondansetron has some utility, with a single dose being associated with less need for intravenous fluids, fewer hospitalizations, and decreased vomiting. Metoclopramide might also be helpful. However, the use of ondansetron might possibly be linked to an increased rate of return to hospital in children. The intravenous preparation of ondansetron may be given orally if clinical judgment warrants. Dimenhydrinate, while reducing vomiting, does not appear to have a significant clinical benefit.

Antibiotics

Antibiotics are not usually used for gastroenteritis, although they are sometimes recommended if symptoms are particularly severe or if a susceptible bacterial cause is isolated or suspected. If antibiotics are to be employed, a macrolide (such as azithromycin) is preferred over a fluoroquinolone due to higher rates of resistance to the latter. Pseudomembranous colitis, usually caused by antibiotic use, is managed by discontinuing the causative agent and treating it with either metronidazole or vancomycin. Bacteria and protozoans that are amenable to treatment include Shigella Salmonella typhi, and Giardia species. In those with Giardia species or Entamoeba histolytica, tinidazole treatment is recommended and superior to metronidazole. The World Health Organization (WHO) recommends the use of antibiotics in young children who have both bloody diarrhea and fever.

Antimotility Agents

Antimotility medication has a theoretical risk of causing complications, and although clinical experience has shown this to be unlikely, these drugs are discouraged in people with bloody diarrhea or diarrhea that is complicated by fever. Loperamide, an opioid analogue, is commonly used for the symptomatic treatment of diarrhea. Loperamide is not recommended in children, however, as it may cross the immature blood–brain barrier and cause toxicity. Bismuth subsalicylate, an insoluble complex of trivalent bismuth and salicylate, can be used in mild to moderate cases, but salicylate toxicity is theoretically possible.

Food Allergy

A food allergy is an abnormal immune response to food. The symptoms of the allergic reaction may range from mild to severe. They may include itchiness, swelling of the tongue, vomiting, diarrhea, hives, trouble breathing, or low blood pressure. This typically occurs within minutes to several hours of exposure. When the symptoms are severe, it is known as anaphylaxis. A food intolerance and food poisoning are separate conditions, not due to an immune response.

Common foods involved include cow's milk, peanuts, eggs, shellfish, fish, tree nuts, soy, wheat, rice, and fruit. The common allergies vary depending on the country. Risk factors include a family history of allergies, vitamin D deficiency, obesity, and high levels of cleanliness. Allergies occur when immunoglobulin E (IgE), part of the body's immune system, binds to food molecules. A protein in the food is usually the problem. This triggers the release of inflammatory chemicals such as histamine. Diagnosis is usually based on a medical history, elimination diet, skin prick test, blood tests for food-specific IgE antibodies, or oral food challenge.

Early exposure to potential allergens may be protective. Management primarily involves avoiding the food in question and having a plan if exposure occurs. This plan may include giving adrenaline (epinephrine) and wearing medical alert jewelry. The benefits of allergen immunotherapy for food allergies is unclear, thus is not recommended as of 2015. Some types of food allergies among children resolve with age, including that to milk, eggs, and soy; while others such as to nuts and shellfish typically do not.

In the developed world, about 4% to 8% of people have at least one food allergy. They are more common in children than adults and appear to be increasing in frequency. Male children appear to be more commonly affected than females. Some allergies more commonly develop early in life, while others typically develop in later life. In developed countries, a large proportion of people believe they have food allergies when they actually do not have them. The declaration of the presence of trace amounts of allergens in foods is mandatory only in Brazil.

Signs and Symptoms

Food allergies usually have a fast onset (from seconds to one hour) and may include:

- Rash,
- Hives,

- Itching of mouth, lips, tongue, throat, eyes, skin, or other areas,
- Swelling (angioedema) of lips, tongue, eyelids, or the whole face,
- Difficulty swallowing,
- Runny or congested nose,
- Hoarse voice,
- Wheezing and shortness of breath,
- Diarrhea, abdominal pain, and stomach cramps,
- Lightheadedness,
- Fainting,
- Nausea,
- Vomiting.

In some cases, however, onset of symptoms may be delayed for hours. Symptoms can vary. The amount of food needed to trigger a reaction also varies.

Serious danger regarding allergies can begin when the respiratory tract or blood circulation is affected. The former can be indicated through wheezing and cyanosis. Poor blood circulation leads to a weak pulse, pale skin and fainting.

A severe case of an allergic reaction, caused by symptoms affecting the respiratory tract and blood circulation, is called anaphylaxis. When symptoms are related to a drop in blood pressure, the person is said to be in anaphylactic shock. Anaphylaxis occurs when IgE antibodies are involved, and areas of the body that are not in direct contact with the food become affected and show symptoms. Those with asthma or an allergy to peanuts, tree nuts, or seafood are at greater risk for anaphylaxis.

Cause

Although sensitivity levels vary by country, the most common food allergies are allergies to milk, eggs, peanuts, tree nuts, seafood, shellfish, soy, and wheat. These are often referred to as "the big eight". Allergies to seeds — especially sesame — seem to be increasing in many countries. An example an allergy more common to a particular region is that to rice in East Asia where it forms a large part of the diet.

One of the most common food allergies is a sensitivity to peanuts, a member of the bean family. Peanut allergies may be severe, but children with peanut allergies sometimes outgrow them. Tree nuts, including cashews, Brazil nuts, hazelnuts, macadamia nuts, pecans, pistachios, pine nuts, coconuts, and walnuts, are also common allergens. Sufferers may be sensitive to one particular tree nut or to many different ones. Also, seeds, including sesame seeds and poppy seeds, contain oils where protein is present, which may elicit an allergic reaction.

Egg allergies affect about one in 50 children but are frequently outgrown by children when they reach age five. Typically, the sensitivity is to proteins in the white, rather than the yolk.

Milk from cows, goats, or sheep is another common food allergen, and many sufferers are also unable to tolerate dairy products such as cheese. A small portion of children with a milk allergy, roughly 10%, have a reaction to beef. Beef contains a small amount of protein that is also present in cow's milk.

Seafood is one of the most common sources of food allergens; people may be allergic to proteins found in fish, crustaceans, or shellfish.

Other foods containing allergenic proteins include soy, wheat, fruits, vegetables, maize, spices, synthetic and natural colors, and chemical additives.

Balsam of Peru, which is in various foods, is in the "top five" allergens most commonly causing patch test reactions in people referred to dermatology clinics.

Sensitization

Sensitization can occur through the gastrointestinal tract, respiratory tract and possibly the skin. Damage to the skin in conditions such as eczema has been proposed as a risk factor for sensitization. An Institute of Medicine report says that food proteins contained in vaccines, such as gelatin, milk, or egg can cause sensitization (development of allergy) in vaccine recipients, to those food items.

Atopy

Food allergies develop more easily in people with the atopic syndrome, a very common combination of diseases: allergic rhinitis and conjunctivitis, eczema, and asthma. The syndrome has a strong inherited component; a family history of allergic diseases can be indicative of the atopic syndrome.

Cross-reactivity

Some children who are allergic to cow's milk protein also show a cross-sensitivity to soy-based products. Some infant formulas have their milk and soy proteins hydrolyzed, so when taken by infants, their immune systems do not recognize the allergen and they can safely consume the product. Hypoallergenic infant formulas can be based on proteins partially predigested to a less antigenic form. Other formulas, based on free amino acids, are the least antigenic and provide complete nutritional support in severe forms of milk allergy.

People with latex allergy often also develop allergies to bananas, kiwifruit, avocados, and some other foods.

Pathophysiology

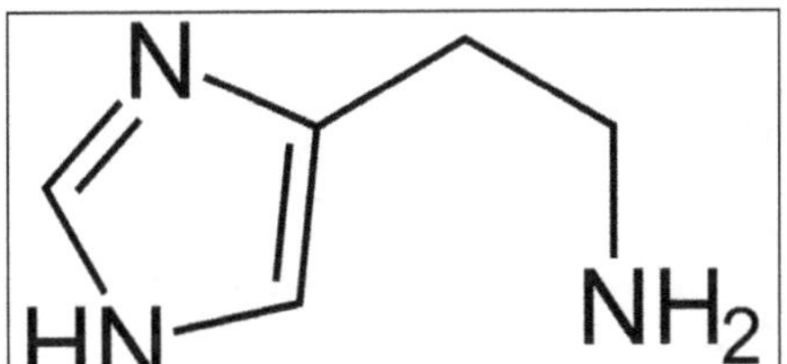

A histamine, the structure shown, causes a person to feel itchy during an allergic reaction. A common medication to stop this is an antihistamine, which fights the histamines in the person's system.

Conditions caused by food allergies are classified into three groups according to the mechanism of the allergic response:

1. IgE-mediated (classic): The most common type, occurs shortly after eating and may involve anaphylaxis.
2. Non-IgE mediated: Characterized by an immune response not involving immunoglobulin E; may occur some hours after eating, complicating diagnosis.
3. IgE and non-IgE-mediated: A hybrid of the above two types.

Allergic reactions are hyperactive responses of the immune system to generally innocuous substances. When immune cells encounter the allergenic protein, IgE antibodies are produced; this is similar to the immune system's reaction to foreign pathogens. The IgE antibodies identify the allergenic proteins as harmful and initiate the allergic reaction. The harmful proteins are those that do not break down due to the strong bonds of the protein. IgE antibodies bind to a receptor on the surface of the protein, creating a tag, just as a virus or parasite becomes tagged. Why some proteins do not denature and subsequently trigger allergic reactions and hypersensitivity while others do is not entirely clear.

Hypersensitivities are categorized according to the parts of the immune system that are attacked and the amount of time it takes for the response to occur. The four types of hypersensitivity reaction are: type 1, immediate IgE-mediated; type 2, cytotoxic; type 3, immune complex-mediated; and type 4, delayed cell-mediated. The pathophysiology of allergic responses can be divided into two phases. The first is an acute response that occurs immediately after exposure to an allergen. This phase can either subside or progress into a "late-phase reaction" which can substantially prolong the symptoms of a response, and result in tissue damage.

Many food allergies are caused by hypersensitivities to particular proteins in different foods. Proteins have unique properties that allow them to become allergens, such as stabilizing forces in their tertiary and quaternary structures which prevent degradation during digestion. Many theoretically allergenic proteins cannot survive the destructive environment of the digestive tract, thus do not trigger hypersensitive reactions.

Acute Response

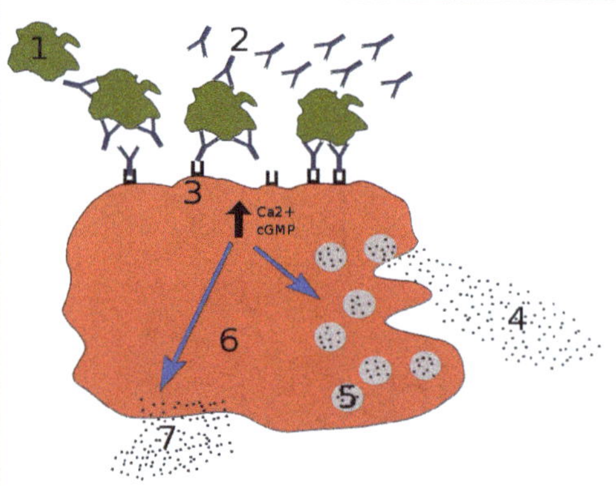

1 — Antigen

2 — Ige antibody

3 — FcεRI receptor

4 — Preformed mediators (histamine, proteases, chemokines, heparin)

5 — Granules

6 — Mast cell

7 — Newly formed mediators (prostaglandins, leukotrienes, thromboxanes, PAF)

Degranulation process in allergy.

In the early stages of allergy, a type I hypersensitivity reaction against an allergen, encountered for the first time, causes a response in a type of immune cell called a T_H2 lymphocyte, which belongs to a subset of T cells that produce a cytokine called interleukin-4 (IL-4). These T_H2 cells interact with other lymphocytes called B cells, whose role is the production of antibodies. Coupled with signals provided by IL-4, this interaction stimulates the B cell to begin production of a large amount of a particular type of antibody known as IgE. Secreted IgE circulates in the blood and binds to an IgE-specific receptor (a kind of Fc receptor called FcεRI) on the surface of other kinds of immune cells called mast cells and basophils, which are both involved in the acute inflammatory response. The IgE-coated cells, at this stage, are sensitized to the allergen.

If later exposure to the same allergen occurs, the allergen can bind to the IgE molecules held on the surface of the mast cells or basophils. Cross-linking of the IgE and Fc receptors occurs when more than one IgE-receptor complex interacts with the same allergenic molecule, and activates the sensitized cell. Activated mast cells and basophils undergo a process called degranulation, during which they release histamine and other inflammatory chemical mediators (cytokines, interleukins, leukotrienes, and prostaglandins) from their granules into the surrounding tissue causing several systemic effects, such as vasodilation, mucous secretion, nerve stimulation, and smooth-muscle contraction. This results in rhinorrhea, itchiness, dyspnea, and anaphylaxis. Depending on the individual, the allergen, and the mode of introduction, the symptoms can be system-wide (classical anaphylaxis), or localized to particular body systems; asthma is localized to the respiratory system and eczema is localized to the dermis.

Late-phase Response

After the chemical mediators of the acute response subside, late-phase responses can often occur due to the migration of other leukocytes such as neutrophils, lymphocytes, eosinophils, and macrophages to the initial site. The reaction is usually seen 2–24 hours after the original reaction. Cytokines from mast cells may also play a role in the persistence of long-term effects. Late-phase responses seen in asthma are slightly different from those seen in other allergic responses, although they are still caused by release of mediators from eosinophils, and are still dependent on activity of T_H2 cells.

Prevention

Breastfeeding for more than four months may prevent atopic dermatitis, cow's milk allergy, and wheezing in early childhood. Early exposure to potential allergens may be protective. Specifically, early exposure to eggs and peanuts reduces the risk of allergies to these. Guidelines suggest introducing peanuts as early as 4–6 months and include precautionary measures for high-risk infants. The former guidelines, advised delaying the introduction of peanuts, are now thought to have contributed to the increase in peanut allergy seen recently.

To avoid an allergic reaction, a strict diet can be followed. It is difficult to determine the amount of allergenic food required to elicit a reaction, so complete avoidance should be attempted. In some cases, hypersensitive reactions can be triggered by exposures to allergens through skin contact, inhalation, kissing, participation in sports, blood transfusions, cosmetics, and alcohol.

Inhalation Exposure

Allergic reactions to airborne particles or vapors of known food allergens have been reported as an occupational consequence of people working in the food industry, but can also take place in home situations, restaurants, or confined spaces such as airplanes. According to two reviews, respiratory symptoms are common, but in some cases there has been progression to anaphylaxis. The most frequent reported cases of reactions by inhalation of allergenic foods were due to peanut, seafood, legumes, tree nut, and cow's milk. Steam rising from cooking of lentils, green beans, chickpeas and fish has been well documented as triggering reactions, including anaphylactic reactions. One review mentioned case study examples of allergic responses to other foods, including examples in which oral consumption of the food is tolerated.

Nausea

Nausea is having a sick or queasy feeling in the stomach, and vomiting is throwing up food or liquids from the stomach. Nausea can happen when a person isn't even thinking about food. And a person can vomit even if they haven't eaten anything. Sometimes they vomit even if they haven't had any nausea.

Nausea or vomiting can be caused by eating something that disagrees with you, by bacteria in food, by infections, or by radiation or chemo treatments for cancer. Many people have little or no nausea or vomiting with these treatments. For others, just thinking about going for one of the treatments can cause nausea or vomiting. If you have nausea just before chemo or other appointments, ask what can be done to lessen this problem. Cancer by itself can also cause nausea and vomiting.

Don't eat your favorite foods if you are not feeling well. If you eat foods you like when you are nauseated, you could find them unappealing when treatment is over because you associate them with feeling sick.

Frequent vomiting can be dangerous because it can lead to dehydration. It can also lead to inhaling food or liquids, which can cause choking and other problems. Talk with your cancer team about what's causing your nausea and vomiting and what you can do about it.

Be sure to tell your cancer care team if you have nausea or are vomiting because there are medicines that can help. These medicines should be taken on a regular schedule, or around the clock, as prescribed by your doctor. And if a certain medicine doesn't work, your cancer care team may be able to recommend another one. It may take a few tries to find the medicines that work best for you.

For Nausea

- If the nausea only happens between meals, keep something in your stomach. Eat frequent, small snacks throughout the day. Snack ideas include smoothies, trail mix, and fruit.
- On chemotherapy treatment days, eat a small meal or snack before treatment.
- Drink clear liquids served cold and sipped slowly. (Clear liquids are those you can see through, such as ginger ale, apple juice, broth, tea, etc).
- Also try popsicles or gelatin. Suck on hard candy with pleasant smells, such as lemon drops or mints, to help get rid of bad tastes. (Don't eat tart candies if you have mouth sores).

- Eat bland foods, such as dry toast and crackers.
- Eat food cold or at room temperature to decrease its smell and taste.
- Avoid fatty, fried, spicy, or very sweet foods.
- Try small amounts of foods high in calories that are easy to eat (such as pudding, ice cream, sherbets, yogurt, and milkshakes) several times a day.
- Use butter, oils, syrups, sauces, and milk in foods to raise calories.
- Avoid low-fat foods unless fats upset your stomach or cause other problems.
- Eat the foods you like. Many people develop a dislike for red meat and meat broths during treatment. Try other protein sources, such as fish, chicken, beans, and nuts.
- Tart or sour foods may be easier to keep down (unless you have mouth sores).
- Try to rest quietly while sitting upright for at least an hour after each meal.
- Distract yourself with soft music, a favourite TV program, or the company of others.
- Tell your cancer team about the nausea, because there are many drugs that can help it.
- Take your anti-nausea medicine at the first signs of nausea to help prevent vomiting.
- While waiting for your nausea medicine to work, relax and take slow, deep breaths.

References

- Elliott, EJ (6 January 2007). "Acute gastroenteritis in children". The BMJ. 334 (7583): 35–40. Doi:10.1136/bmj.39036.406169.80. PMC 1764079. PMID 17204802
- Basem Abdelmalak, John Doyle, eds. (2013). Anesthesia for otolaryngologic surgery. Cambridge University Press. Pp. 282–287. ISBN 978-1-107-01867-9
- Four-most-common-types-of-food-contamination: kent.co.in, Retrieved 29 March, 2019
- "Diarrhoeal disease Fact sheet N°330". World Health Organization. April 2013. Archivedfrom the original on 17 July 2014. Retrieved 9 July 2014
- Alam NH, Ashraf H (2003). "Treatment of infectious diarrhea in children". Paediatric Drugs. 5 (3): 151–65. Doi:10.2165/00128072-200305030-00002. PMID 12608880
- Athas, Eric (28 January 2019). "Which Allergens Are in Your Food? You Can't Always Tell From the Labels". The New York Times. ISSN 0362-4331. Retrieved 9 December 2019
- Nausea-and-vomiting/nausea-and-vomiting, treatments-and-side-effects, treatment: cancer.org, Retrieved 30 April, 201

7
Food Allergies

An abnormal immune response to food is known as a food allergy. A few of such food allergies include corn allergy, egg allergy, fruit allergy, milk allergy, soy allergy, wheat allergy and peanut allergy. This chapter has been carefully written to provide an easy understanding of these kinds of food allergies.

Corn Allergy

Corn allergy isn't as common as some allergic reactions to foods, but when it occurs, it can be severe. Symptoms can range from itching, redness, and nasal congestion, to wheezing, throat swelling, and shock (anaphylaxis).

Getting a corn allergy diagnosis can be challenging and one is usually based on history. The only treatment for anaphylaxis is epinephrine injection (EpiPen), making prevention the primary objective.

While many corn-containing foods are obvious (cornstarch, popcorn), others may not be. Corn can even hide in everyday items that have nothing to do with your diet.

Signs and Symptoms

Allergic reactions to corn can take different forms. Common symptoms include:

- Hives,
- Itching, particularly in or around the mouth (oral allergy syndrome), but may be generalized as well,

- Flushing or reddening of the skin,
- Hay fever-like symptoms: sneezing, nasal congestion, runny nose,
- Wheezing, asthma,
- Headaches,
- Abdominal pain,
- Nausea and vomiting,
- Diarrhea.

Anaphylaxis may also occur and can include symptoms such as:

- Swelling and tightness of the lips, tongue, throat, neck, or face,
- Difficulty breathing,
- Hoarse voice,
- Lightheadedness,
- Rapid heart rate,
- Lethargy, confusion, or loss of consciousness,
- A sense of impending doom.

Cause

Unlike some other food allergies (nuts, shellfish, and wheat, for example), the exact causes of corn allergy aren't known. It's thought that a combination of genetic factors, environmental factors, and epigenetic factors (the interaction of genetics and the environment) are at play.

Corn is a cereal grain that contains a protein, zein, which is the suspected culprit in this allergy. A reaction occurs when the body recognizes this protein as foreign and releases immunoglobulin E (IgE), antibodies to attack the protein. IgE then stimulates cells in the immune system to secrete substances such as histamines that are responsible for the symptoms.

Allergic reactions can occur as a result of eating both raw and cooked corn, as well as foods manufactured with corn products. Not all corn products contain zein, but it can be difficult to know when it is present, as current food labeling does include a "corn free" designation.

Even coming into contact with surgical gloves or intravenous fluids that contain corn can be problematic. Those who have corn allergy may also react to corn pollen, grass pollen, and cornstarch (typically with hay fever (allergic rhinitis) and asthma).

People who have asthma, eczema, hives, hay fever, or other food allergies appear to be at greater risk. A family history of these conditions is also associated with a higher risk, particularly when a sibling has a corn allergy.

Diagnosis and Testing

Diagnosing food allergies is important, as the results can have a significant impact on what a person eats each and every day. That said, the diagnosis of corn allergy can be challenging.

Allergy testing with blood tests and skin tests can be inaccurate, with false positive tests occurring often. Such a result, however, does place a person at a higher risk for an allergic reaction to that food and should be considered along with other findings.

A careful history is often the most reliable indicator of a corn allergy, with symptoms of an allergic reaction occurring after eating corn or foods containing corn. The history, however, can be difficult to evaluate for a few reasons:

- Corn is present in a vast number of foods and in different amounts.
- Signs and symptoms are non-specific and may easily be dismissed as a cold virus, a rash due to irritation, or an allergic reaction due to something else.

With mild symptoms, keeping a food diary is often an excellent start. This involves recording foods that are eaten, when they are eaten, and any symptoms you experience.

An elimination diet can also be very helpful. With this, the foods that are eaten are greatly restricted, and then individual foods are slowly added back in at specific intervals so that a reaction, if present, can easily be traced back.

The diet often requires a commitment of a minimum of two weeks and often more to identify potential food allergies. If a corn allergy is suspected, a food challenge (eating corn) may be considered, but should only be done under the guidance of an allergist.

Consulting with an allergist who specializes in food allergies early on can be very helpful, and is imperative if you have had any symptoms suggestive of an anaphylactic reaction.

Treatment

For mild allergic reactions, treatment usually consists of managing the symptoms alone until the allergic reaction is done.

For anaphylactic reactions, epinephrine (an EpiPen) is the only treatment available, along with immediate medical care in an emergency room.

Prevention

Certainly, the best way to "treat" corn allergy is to avoid corn in the first place. That said, with the many hidden sources of corn in a typical American diet, and the lack of clear labeling to this end, exposures can and do occur.

Working with a nutritionist or dietitian can be very helpful in navigating food label vocabulary that indicates the presence of corn in a product. But even when you become familiar with how to identify these ingredients, it can be challenging when you are asked to eat at the homes of family and friends, out at restaurants, and for children, at school.

At restaurants, it's a good idea to ask to talk with the chef. This conversation should include not only the ingredients used but how the food is prepared (for example, a corn-free dish may be fried in corn oil).

Egg Allergy

Egg allergy is an immune hypersensitivity to proteins found in chicken eggs, and possibly goose, duck, or turkey eggs. Symptoms can be either rapid or gradual in onset. The latter can take hours to days to appear. The former may include anaphylaxis, a potentially life-threatening condition which requires treatment with epinephrine. Other presentations may include atopic dermatitis or inflammation of the esophagus.

In the United States, 90% of allergic responses to foods are caused by cow's milk, eggs, wheat, shellfish, peanuts, tree nuts, fish, and soy beans. The declaration of the presence of trace amounts of allergens in foods is not mandatory in any country, with the exception of Brazil.

Prevention is by avoiding eating eggs and foods that may contain eggs, such as cake or cookies. It is unclear if the early introduction of the eggs to the diet of babies aged 4–6 months decreases the risk of egg allergies.

Egg allergy appears mainly in children but can persist into adulthood. In the United States, it is the second most common food allergy in children after cow's milk. Most children outgrow egg allergy by the age of five, but some people remain allergic for a lifetime. In North America and Western Europe egg allergy occurs in 0.5% to 2.5% of children under the age of five years. The majority grow out of it by school age, but for roughly one-third, the allergy persists into adulthood. Strong predictors for adult-persistence are anaphylaxis, high egg-specific serum immunoglobulin E (IgE), robust response to the skin prick test and absence of tolerance to egg-containing baked foods.

Signs and Symptoms

Food allergies usually have a fast onset (from seconds to one hour). Symptoms may include: rash, hives, itching of mouth, lips, tongue, throat, eyes, skin, or other areas, swelling of lips, tongue, eyelids, or the whole face, difficulty swallowing, runny or congested nose, hoarse voice, wheezing, shortness of breath, diarrhea, abdominal pain, lightheadedness, fainting, nausea, or vomiting. Symptoms of allergies vary from person to person and may vary from incident to incident. Serious danger regarding allergies can begin when the respiratory tract or blood circulation is affected. The former can be indicated by wheezing, a blocked airway and cyanosis, the latter by weak pulse, pale skin, and fainting. When these symptoms occur the allergic reaction is called anaphylaxis. Anaphylaxis occurs when IgE antibodies are involved, and areas of the body that are not in direct contact with the food become affected and show severe symptoms. Untreated, this can proceed to vasodilation, a low blood pressure situation called anaphylactic shock, and death (very rare).

Young children may exhibit dermatitis/eczema on face, scalp and other parts of the body, in older children knees and elbows are more commonly afflicted. Children with dermatitis are at greater than expected risk of also exhibiting asthma and allergic rhinitis.

Causes

Eating Egg

The cause is typically the eating of eggs or foods that contain eggs. Briefly, the immune system over-reacts to proteins found in eggs. This allergic reaction may be triggered by small amounts of egg, even egg incorporated into cooked foods, such as cake. People with an allergy to chicken eggs may also be reactive to goose, duck, or turkey eggs.

Vaccines

Influenza vaccines are created by injecting a live virus into fertilized chicken eggs. The viruses are harvested, killed and purified, but a residual amount of egg white protein remains. Each year, vaccines are created to provide protection against the flu viruses expected to be prevalent in the upcoming cold weather months. For the 2017-2018 flu season, the vaccines are described as IIV3 and IIV4 for resistance to the expected three or four viruses. For adults ages 18 and older there is also an option to receive recombinant flu vaccines (RIV3 or RIV4) which are grown on mammalian cell cultures instead of in eggs, and so are no risk for people with severe egg allergy. Recommendations are that for people with a history of mild egg allergy should receive any IIV or RIV vaccine. People with a more severe allergic reaction may also receive any IIV or RIV, but in an inpatient or outpatient medical setting, administered by a healthcare provider. People with a known severe allergic reaction to influenza vaccine (which could be egg protein or the gelatin or the neomycin components of the vaccine) should not receive a flu vaccine.

Each year the American Academy of Pediatrics (AAP) publishes recommendations for prevention and control of influenza in children. In the most recent guidelines, for 2016-2017, a change was made, that children with a history of egg allergy may receive the IIV3 or IIV4 vaccine without special precautions. It does, however, state that "Standard vaccination practice should include the ability to respond to acute hypersensitivity reactions." Prior to this, AAP recommended precautions based on egg allergy history: if no history, immunize; if a history of mild reaction, i.e., hives, immunize in a medical setting with healthcare professionals and resuscitative equipment available; if a history of severe reactions, refer to an allergist.

The measles and mumps parts of the "MMR vaccine" (for measles, mumps, and rubella) are cultured on chick embryo cell culture and contain trace amounts of egg protein. The amount of egg protein is lower than in influenza vaccines and the risk of an allergic reaction is much lower. One guideline stated that all infants and children should get the two MMR vaccinations, mentioning that "Studies on large numbers of egg-allergic children show there is no increased risk of severe allergic reactions to the vaccines." Another guideline recommended that if a child has a known medical history of severe anaphylaxis reaction to eggs, then the vaccination should be done in a hospital center, and the child be kept for observation for 60 minutes before being allowed to leave. The second guideline also stated that if there was a severe reaction to the first vaccination - which could have been to egg protein or the gelatin and neomycin components of the vaccine - the second is contraindicated.

Exercise

There is a condition called food-dependent, exercise-induced anaphylaxis (FDEIAn). Exercise can trigger hives and more severe symptoms of an allergic reaction. For some people with this condition, exercise alone is not sufficient, nor consumption of a food to which they are mildly allergic sufficient, but when the food in question is consumed within a few hours before high intensity exercise, the result can be anaphylaxis. Egg are specifically mentioned as a causative food. One theory is that exercise is stimulating the release of mediators such as histamine from IgE-activated mast cells. Two of the reviews postulate that exercise is not essential for the development of symptoms, but rather that it is one of several augmentation factors, citing evidence that the culprit food in combination with alcohol or aspirin will result in a respiratory anaphylactic reaction.

Mechanisms

Conditions caused by food allergies are classified into three groups according to the mechanism of the allergic response:

1. IgE-mediated (classic): The most common type, manifesting acute changes that occur shortly after eating, and may progress to anaphylaxis.
2. Non-IgE mediated: Characterized by an immune response not involving immunoglobulin E; may occur hours to days after eating, complicating diagnosis.
3. IgE and non-IgE-mediated: A hybrid of the above two types.

Allergic reactions are hyperactive responses of the immune system to generally innocuous substances, such as proteins in the foods we eat. Why some proteins trigger allergic reactions while others do not is not entirely clear, although in part thought to be due to resistance to digestion. Because of this, intact or largely intact proteins reach the small intestine, which has a large presence of white blood cells involved in immune reactions. The heat of cooking structurally degrades protein molecules, potentially making them less allergenic. The pathophysiology of allergic responses can be divided into two phases. The first is an acute response that occurs immediately after exposure to an allergen. This phase can either subside or progress into a "late-phase reaction" which can substantially prolong the symptoms of a response, and result in more tissue damage.

In the early stages of acute allergic reaction, lymphocytes previously sensitized to a specific protein or protein fraction react by quickly producing a particular type of antibody known as secreted IgE (sIgE), which circulates in the blood and binds to IgE-specific receptors on the surface of other kinds of immune cells called mast cells and basophils. Both of these are involved in the acute inflammatory response. Activated mast cells and basophils undergo a process called degranulation, during which they release histamine and other inflammatory chemical mediators called (cytokines, interleukins, leukotrienes, and prostaglandins) into the surrounding tissue causing several systemic effects, such as vasodilation, mucous secretion, nerve stimulation, and smooth-muscle contraction. This results in runny nose, itchiness, shortness of breath, and potentially anaphylaxis. Depending on the individual, the allergen, and the mode of introduction, the symptoms can be system-wide (classical anaphylaxis), or localized to particular body systems; asthma is localized to the respiratory system while eczema is localized to the skin.

After the chemical mediators of the acute response subside, late-phase responses can often occur due to the migration of other white blood cells such as neutrophils, lymphocytes, eosinophils, and macrophages to the initial reaction sites. This is usually seen 2–24 hours after the original reaction. Cytokines from mast cells may also play a role in the persistence of long-term effects.

Late-phase responses seen in asthma are slightly different from those seen in other allergic responses, although they are still caused by release of mediators from eosinophils.

Five major allergenic proteins from the egg of the domestic chicken (Gallus domesticus) have been identified; these are designated Gal d 1-5. Four of these are in egg white: ovomucoid (Gal d 1), ovalbumin (Gal d 2), ovotransferrin (Gal d 3) and lysozyme (Gal d 4). Of these, ovomucoid is the dominant allergen, and one that is less likely to be outgrown as children get older. Ingestion of under-cooked egg may trigger more severe clinical reactions than well-cooked egg. In egg yolk, alpha-livetin (Gal d 5) is the major allergen, but various vitellins may also trigger a reaction. People allergic to alpha-livetin may experience respiratory symptoms such as rhinitis and asthma when exposed to chickens, because the yolk protein is also found in live birds. In addition to IgE-mediated responses, egg allergy can manifest as atopic dermatitis, especially in infants and young children. Some will display both, so that a child could react to an oral food challenge with allergic symptoms, followed a day or two later with a flare up of atopic dermatitis and gastrointestinal symptoms, including allergic eosinophilic esophagitis.

Non-allergic Intolerance

Egg whites, which are potentially histamine liberators, also provoke a nonallergic response in some people. In this situation, proteins in egg white directly trigger the release of histamine from mast cells. Because this mechanism is classified as a pharmacological reaction, or "pseudoallergy", the condition is considered a food intolerance instead of a true immunoglobulin E (IgE) based allergic reaction.

The response is usually localized, typically in the gastrointestinal tract. Symptoms may include abdominal pain, diarrhea, or any other symptoms typical to histamine release. If sufficiently strong, it can result in an anaphylactoid reaction, which is clinically indistinguishable from true anaphylaxis. Some people with this condition tolerate small quantities of egg whites. They are more often able to tolerate well-cooked eggs, such as found in cake or dried egg-based pasta, than incompletely cooked eggs, such as fried eggs or meringues, or uncooked eggs.

Diagnosis

Diagnosis of egg allergy is based on the person's history of allergic reactions, skin prick test (SPT), patch test and measurement of egg-specific serum immunoglobulin E (IgE or sIgE). Confirmation is by double-blind, placebo-controlled food challenges. SPT and sIgE have sensitivity greater than 90% but specificity in the 50-60% range, meaning these tests will detect an egg sensitivity, but will also be positive for other allergens. For young children, attempts have been made to identify SPT and sIgE responses strong enough to avoid the need for a confirming oral food challenge.

Prevention

When eggs are introduced to a baby's diet is thought to affect risk of developing allergy, but there are contradictory recommendations. A 2016 review acknowledged that introducing peanuts early appears to have a benefit, but stated "The effect of early introduction of egg on egg allergy are controversial." A meta-analysis published the same year supported the theory that early introduction of eggs into an infant's diet lowers risk, and a review of allergens in general stated that introducing solid foods at 4–6 months may result in the lowest subsequent allergy risk. However, an older consensus document from the American College of Allergy, Asthma and Immunology recommended that introduction of chicken eggs be delayed to 24 months of age.

Treatment

The mainstay of treatment is total avoidance of egg protein intake. This is complicated because the declaration of the presence of trace amounts of allergens in foods is not mandatory.

Treatment for accidental ingestion of egg products by allergic individuals varies depending on the sensitivity of the person. An antihistamine such as diphenhydramine (Benadryl) may be prescribed. Sometimes prednisone will be prescribed to prevent a possible late phase Type I hypersensitivity reaction. Severe allergic reactions (anaphalaxis) may require treatment with an epinephrine pen, i.e., an injection device designed to be used by a non-healthcare professional when emergency treatment is warranted. A second dose is needed in 16-35% of episodes.

Immunotherapy

There is active research on trying oral immunotherapy (OIT) to desensitize people to egg allergens. A Cochrane Review concluded that OIT can desensitize people, but it remains unclear whether long-term tolerance develops after treatment ceases, and 69% of the people enrolled in the trials had adverse effects. They concluded there was a need for standardized protocols and guidelines prior to incorporating OIT into clinical practice. A second review noted that allergic reactions, up to anaphylaxis, can occur during OIT, and recommends this treatment not be routine medical practice. A third review limited its scope to trials of baked egg-containing goods such as bread or cake as a means of resolving egg allergy. Again, there were some successes, but also some severe allergic reactions, and the authors came down on the side of not recommending this as treatment.

Avoiding Eggs

Prevention of egg-allergic reactions means avoiding eggs and egg-containing foods. People with an allergy to chicken eggs may also be allergic to other types of eggs, such as goose, duck, or turkey eggs. In cooking, eggs are multifunctional: they may act as an emulsifier to reduce oil/water separation (mayonnaise), a binder (water binding and particle adhesion, as in meatloaf), or an aerator (cakes, especially angel food). Some commercial egg substitutes can substitute for particular functions (potato starch and tapioca for water binding, whey protein for aeration or particle binding, or soy lecithin or avocado for emulsification). Food companies produce egg-free mayonnaise and other replacement foods. Alfred Bird invented egg-free Bird's Custard, the original version of what is known generically as custard powder today.

Most people find it necessary to strictly avoid any item containing eggs, including:

- Albumin (egg white protein),
- Apovitellin (egg yolk protein),
- Egg Beaters (cholesterol-free, uses egg whites),

- Dried egg solids, powdered egg,
- Egg, egg white, egg yolk,
- Egg wash,
- Eggnog,
- Fat substitutes (some),
- Livetin (egg yolk protein),
- Lysozyme (egg white protein),
- Mayonnaise,
- Meringue or meringue powder,
- Ovalbumin (egg white protein),
- Ovoglobulin (egg white protein),
- Ovomucin (egg white protein),
- Ovomucoid (egg white protein),
- Ovotransferrin (egg white protein),
- Ovovitelia (egg yolk protein),
- Ovovitellin (egg yolk protein),
- Silici albuminate,
- Simplesse,
- Vitellin (egg yolk protein).

Ingredients that sometimes include egg protein include: artificial flavoring, natural flavoring, lecithin and nougat candy.

Probiotic products have been tested, and some found to contain milk and egg proteins which were not always indicated on the labels.

Prognosis

The majority of children outgrow egg allergy. One review reported that 70% of children will outgrow this allergy by 16 years. In subsequently published longitudinal studies, one reported that for 140 infants who had challenge-confirmed egg allergy, 44% had resolved by two years. A second reported that for 203 infants with confirmed IgE-mediated egg allergy, 45% resolved by two years of age, 66% by four years, and 71% by six years. Children will be able to tolerate eggs as an ingredient in baked goods and well-cooked eggs sooner than under-cooked eggs. Resolution was more likely if baseline serum IgE was lower, and if the baseline symptoms did not include anaphylaxis.

Fruit Allergy

A fruit allergy is a food allergy. Fruit allergies make up about ten percent of all food related allergies.

Symptoms

Allergic reactions to fruit and vegetables are usually mild and often just affect the mouth, causing itching, a rash, or blisters where the food touches the lips and mouth. This is called oral allergy syndrome. A number of people who react in this way to fruit or vegetables will also react to pollen from some trees and weeds. So, for example, people who are allergic to birch pollen are also likely to be allergic to apples. Another symptom may include slight swelling in the throat, making it feel like it is closing. The ability to breathe is still present though, so it is not fatal.

Other Symptoms due to Hypersensitivity

The symptoms may vary depending upon the person, the severity of the allergy, and type of fruit. For example, mango allergy symptoms include hoarseness, dyspnoea and bronchitic rales (asthma) (Sareen and Shah). The duration of the symptoms tested by Saree and Shah were variable and ranged from 4 h to 7 days. The symptoms may appear within a few minutes.

Different Allergic Fruits

There are many different types of fruits that people have been shown to react allergically such as mangoes and bananas. Some foods are clearly more allergenic than others. In adults, peanuts, tree nuts, finned fish, crustaceans, fruit, and vegetables account for 85% of the food-allergic reactions (O'Neil, Zanovec and Nickla). People suffering from allergies may suffer from a hypersensitivity to the allergic food, which is what causes the allergic reaction. Most fruit allergies are oral syndrome allergies because they are consumed but may also be an external allergy if the fruit touches the skin.

Diagnostic

Skin prick testing is a common way of testing for an allergy. Other ways to test for allergies can be challenge testing, which consists in feeding a very small and measured amount of the allergen to the patient and monitor the reaction (O'Neil, Zanovec and Nickla). This should only be done by a doctor under surveillance.

Allergy or Intolerance

An allergy is different from an intolerance. Food allergies and food intolerances should not be confused because they do not contain the same risks and are not diagnosed the same way. Allergies can be fatal after only a small consumption, while intolerance, although uncomfortable, are not as deadly. An intolerance may lead to a nutrient deficiency which could cause death if untreated but the intolerance itself is not enough to cause rapid death. Allergies, with their varying symptoms, could cause instantaneous death if there is inflammation in the throat and causes suffocation.

Mitigation

For those allergic to fruits, cooking may help reduce or eliminate the reaction to some fruits. People with this allergy might not necessarily be allergic to citrus fruits.

Milk Allergy

Milk allergy is an adverse immune reaction to one or more proteins in cow's milk. When allergy symptoms occur, they can occur rapidly or have a gradual onset. The former may include anaphylaxis, a potentially life-threatening condition which requires treatment with epinephrine among other measures. The latter can take hours to days to appear, with symptoms including atopic dermatitis, inflammation of the esophagus, enteropathy involving the small intestine and proctocolitis involving the rectum and colon.

In the United States, 90% of allergic responses to foods are caused by eight foods, with cow's milk being the most common. Recognition that a small number of foods are responsible for the majority of food allergies has led to requirements to prominently list these common allergens, including dairy, on food labels. One function of the immune system is to defend against infections by recognizing foreign proteins. It should not over-react to food proteins. Stomach acids cause most proteins to become denatured, meaning to lose their 3-dimensional configuration, and thus lose allergenicity. Heat via cooking can have the same effect. Immune tolerance is another safeguard to not over-reacting to food proteins.

Management is by avoiding eating any dairy foods or foods that contain dairy ingredients. In people with rapid reactions (IgE-mediated milk allergy), the dose capable of provoking an allergic response can be as low as a few milligrams, so recommendations are to avoid dairy strictly. The declaration of the presence of trace amounts of milk or dairy in foods is not mandatory in any country, with the exception of Brazil.

Milk allergy affects between 2% and 3% of babies and young children. To reduce risk, recommendations are that babies should be exclusively breastfed for at least four months, preferably six months, before introducing cow's milk. If there is a family history of dairy allergy, then soy infant formula can be considered, but about 10 to 15% of babies allergic to cow's milk will also react to soy. The majority of children outgrow milk allergy, but for about 0.4% the condition persists into adulthood. Oral immunotherapy is being researched, but it is of unclear benefit.

Signs and Symptoms

Food allergies can have rapid-onset (from minutes up to 2 hours), delayed-onset (up to 48 hours or even 1 week), or combinations of both, depending on the mechanisms involved. The difference depends on the types of white blood cells involved. B cells, a subset of white blood cells, rapidly synthesize and secrete immunoglobulin E (IgE), a class of antibody which bind to antigens, i.e., the foreign proteins. Thus, immediate reactions are described as IgE-mediated. The delayed reactions involve non-IgE-mediated immune mechanisms initiated by B cells, T cells, and other white blood cells. Unlike with IgE reactions, there are no specific biomarker molecules circulating in the blood, and so, confirmation is by removing the suspect food from the diet and see if the symptoms resolve.

IgE-mediated symptoms include: rash, hives, itching of the mouth, lips, tongue, throat, eyes, skin, or other areas, swelling of the lips, tongue, eyelids, or the whole face, difficulty swallowing, runny or congested nose, hoarse voice, wheezing, shortness of breath, diarrhea, abdominal pain, light-headedness, fainting, nausea and vomiting. Symptoms of allergies vary from person to person and may also vary from incident to incident. Serious danger regarding allergies can begin when the respiratory tract or blood circulation is affected. The former can be indicated by wheezing, a blocked airway and cyanosis, the latter by weak pulse, pale skin, and fainting. When these symptoms occur, the allergic reaction is called anaphylaxis. Anaphylaxis occurs when IgE antibodies are involved, and areas of the body that are not in direct contact with the food become affected and show severe symptoms. Untreated, this can proceed to vasodilation, a low blood pressure situation called anaphylactic shock, and very rarely, death.

For milk allergy, non-IgE-mediated responses are more common than IgE-mediated. The presence of certain symptoms, such as angioedema or atopic eczema, is more likely related to IgE-mediated allergies, whereas non-IgE-mediated reactions manifest as gastrointestinal symptoms, without skin or respiratory symptoms. Within non-IgE cow's milk allergy, clinicians distinguish among food protein-induced enterocolitis syndrome (FPIES), food protein-induced allergic proctocolitis (FPIAP) and food protein-induced enteropathy (FPE). Common trigger foods for all are cow's milk and soy foods (including soy infant formula). FPIAP is considered to be at the milder end of the spectrum, and is characterized by intermittent bloody stools. FPE is identified by chronic diarrhea which will resolve when the offending food is removed from the infant's diet. FPIES can be severe, characterized by persistent vomiting, 1 to 4 hours after an allergen-containing food is ingested, to the point of lethargy. Watery and sometimes bloody diarrhea can develop 5 to 10 hours after the triggering meal, to the point of dehydration and low blood pressure. Infants reacting to cow's milk may also react to soy formula, and vice versa. International consensus guidelines have been established for the diagnosis and treatment of FPIES.

Mechanisms

Conditions caused by food allergies are classified into three groups according to the mechanism of the allergic response:

1. IgE-mediated (classic): The most common type, manifesting as acute changes that occur shortly after eating, and may progress to anaphylaxis.
2. Non-IgE mediated: Characterized by an immune response not involving IgE; may occur hours to days after eating, complicating the diagnosis.
3. IgE- and non-IgE-mediated: A hybrid of the above two types.

Allergic reactions are hyperactive responses of the immune system to generally innocuous substances, such as proteins in the foods we eat. Some proteins trigger allergic reactions while others do not. One theory is resistance to digestion, the thinking being that when largely intact proteins reach the small intestine the white blood cells involved in immune reactions will be activated. The heat of cooking structurally degrades protein molecules, potentially making them less allergenic.

Allergic responses can be divided into two phases: An acute response that occurs immediately after exposure to an allergen, which can then either subside or progress into a "late-phase reaction," prolonging the symptoms of a response and resulting in more tissue damage.

In the early stages of acute allergic reaction, lymphocytes previously sensitized to a specific protein or protein fraction react by quickly producing a particular type of antibody known as secreted IgE (sIgE), which circulates in the blood and binds to IgE-specific receptors on the surface of other kinds of immune cells called mast cells and basophils. Both of these are involved in the acute inflammatory response. Activated mast cells and basophils undergo a process called degranulation, during which they release histamine and other inflammatory chemical mediators (cytokines, interleukins, leukotrienes, and prostaglandins) into the surrounding tissue causing several systemic effects, such as vasodilation, mucous secretion, nerve stimulation, and smooth muscle contraction. This results in runny nose, itchiness, shortness of breath, and potentially anaphylaxis. Depending on the individual, the allergen, and the mode of introduction, the symptoms can be system-wide (classical anaphylaxis), or localized to particular body systems; asthma is localized to the respiratory system, while eczema is localized to the skin.

After the chemical mediators of the acute response subside, late-phase responses can often occur due to the migration of other white blood cells such as neutrophils, lymphocytes, eosinophils, and macrophages to the initial reaction sites. This is usually seen 2–24 hours after the original reaction. Cytokines from mast cells may also play a role in the persistence of long-term effects. Late-phase responses seen in asthma are slightly different from those seen in other allergic responses, although they are still caused by release of mediators from eosinophils.

Six major allergenic proteins from cow's milk have been identified: αs1-, αs2-, β-, and κ-casein from casein proteins and α-lactalbumin and β-lactoglobulin from whey proteins. There is some cross-reactivity with soy protein, particularly in non-IgE mediated allergy. Heat can reduce allergenic potential, so dairy ingredients in baked goods may be less likely to trigger a reaction than milk or cheese. For milk allergy, non-IgE-mediated responses are more common than IgE-mediated. The former can manifest as atopic dermatitis and gastrointestinal symptoms, especially in infants and young children. Some will display both, so that a child could react to an oral food challenge with respiratory symptoms and hives (skin rash), followed a day or two later with a flare up of atopic dermatitis and gastrointestinal symptoms, including chronic diarrhea, blood in the stools, gastroesophageal reflux disease (GERD), constipation, chronic vomiting and colic.

Diagnosis

Diagnosis of milk allergy is based on the person's history of allergic reactions, skin prick test (SPT), patch test, and measurement of milk protein specific serum IgE. A negative IgE test does not rule out non-IgE-mediated allergy, also described as cell-mediated allergy. Confirmation is by double-blind, placebo-controlled food challenges, conducted by an allergy specialist. SPT and IgE have a sensitivity of around 88% but specificity of 68% and 48%, respectively, meaning these tests will probably detect a milk sensitivity but may also be false-positive for other allergens.

Attempts have been made to identify SPT and IgE responses accurate enough to avoid the need for confirmation with an oral food challenge. A systematic review stated that in children younger than two years, cut-offs for specific IgE or SPT seem to be more homogeneous and may be proposed.

For older children, the tests were less consistent. It concluded "None of the cut-offs proposed in the literature can be used to definitely confirm cow's milk allergy diagnosis, either to fresh pasteurized or to baked milk."

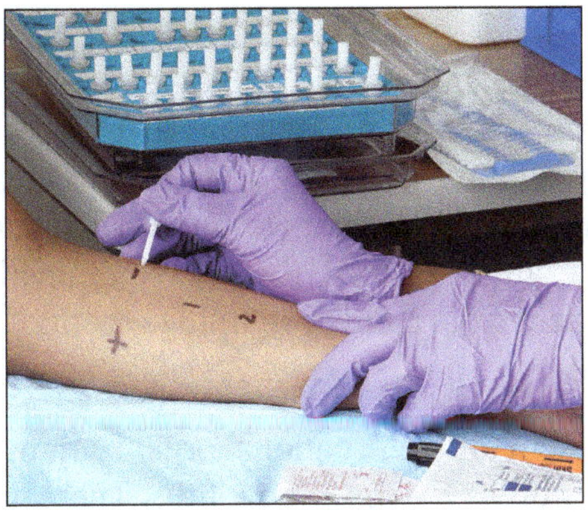

Skin prick testing for allergies. For a positive response, the skin will become red and raised.

Differential Diagnosis

The symptoms of milk allergy can be confused with other disorders that present similar clinical features, such as lactose intolerance, infectious gastroenteritis, celiac disease, non-celiac gluten sensitivity, inflammatory bowel disease, eosinophilic gastroenteritis, and pancreatic insufficiency, among others.

Lactose Intolerance

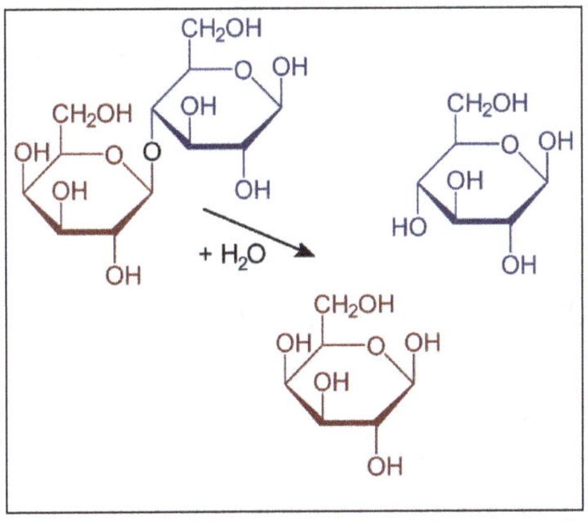

Hydrolysis of the disaccharide lactose to glucose and galactose.

Milk allergy is distinct from lactose intolerance, which is a nonallergic food sensitivity, due to the lack of the enzyme lactase in the small intestines to break lactose down into glucose and galactose. The unabsorbed lactose reaches the large intestine, where resident bacteria use it for fuel, releasing hydrogen, carbon dioxide and methane gases. These gases are the cause of abdominal pain and other symptoms. Lactose intolerance does not cause damage to the gastrointestinal tract. There are four types: primary, secondary, developmental, and congenital. Primary lactose intolerance is when the amount of lactase declines as people age. Secondary lactose intolerance is due to injury to the small intestine, such as from infection, celiac disease, inflammatory bowel disease, or other diseases. Developmental lactose intolerance may occur in premature babies and usually improves over a short period of time. Congenital lactose intolerance is an extremely rare genetic disorder in which little or no lactase is made from birth.

Prevention

Research on prevention addresses the question of whether it is possible to reduce the risk of developing an allergy in the first place. Reviews concluded that there is no strong evidence to recommend changes to the diets of pregnant or nursing women as a means of preventing the development of food allergy in their infants. For mothers of infants considered at high risk of developing cow's milk allergy because of a family history, there is some evidence that the nursing mother avoiding allergens may reduce risk of the child developing eczema, but a Cochrane review concluded that more research is needed.

Guidelines from various government and international organizations recommend that for the lowest allergy risk, infants be exclusively breastfed for 4–6 months. There does not appear to be any benefit to extending that period beyond six months. If a nursing mother decides to start feeding with an infant formula prior to four months the recommendation is to use a formula containing cow's milk proteins.

A different consideration occurs when there is a family history – either parents or older siblings – of milk allergy. The three options to avoiding formula with intact cow's milk proteins are substituting a product containing either extensively hydrolyzed milk proteins, or a non-dairy formula, or one utilizing free amino acids. The hydrolyzation process breaks intact proteins into fragments, in theory reducing allergenic potential. In 2016, the U.S. Food and Drug Administration (FDA) approved a label claim for hydrolyzed whey protein being hypoallergenic. However, a meta-analysis published the same year disputed this claim, concluding that, based on dozens of clinical trials, there was insufficient evidence to support a claim that a partially hydrolyzed formula could reduce the risk of eczema. Soy formula is a common substitution, but infants with milk allergy may also have an allergic response to soy formula. Hydrolyzed rice formula is an option, as are the more expensive amino acid-based formulas.

Treatment

The need for a dairy-free diet should be reevaluated every six months by testing milk-containing products low on the "milk ladder", such as fully cooked, i.e., baked foods, containing milk, in which the milk proteins have been denatured, and ending with fresh cheese and milk. Desensitization via oral immunotherapy is considered experimental.

Treatment for accidental ingestion of milk products by allergic individuals varies depending on the sensitivity of the person. An antihistamine, such as diphenhydramine (Benadryl), may be prescribed. Sometimes prednisone will be prescribed to prevent a possible late-phase type I hypersensitivity reaction. Severe allergic reactions (anaphylaxis) may require treatment with an epinephrine pen, i.e., an injection device designed to be used by a non-healthcare professional when emergency treatment is warranted. A second dose is needed in 16–35% of episodes.

Avoiding Dairy

Most people find it necessary to strictly avoid any item containing dairy ingredients. The reason is that the individual threshold dose capable of provoking an allergic reaction can be quite small, especially in infants. An estimated 5% react to less than 30 milligrams of dairy proteins, and 1% react to less than 1 milligram. A more recent review calculated that the eliciting threshold dose for an allergic reaction in 1% of people (ED01) with confirmed cow's milk allergy is 0.1 mg of cow's milk protein.

Beyond the obvious (anything with milk, cheese, cream, curd, butter, ghee or yogurt in the name), in countries where allergen labeling is mandatory, the ingredient list is supposed to list all ingredients. Anyone with or caring for a person with a dairy protein allergy should always carefully read food package labels, as sometimes even a familiar brand undergoes an ingredient change. In the U.S., for all foods except meat, poultry and egg processed products and most alcoholic beverages, if an ingredient is derived from one of the required-label allergens, then it must either have the food name in parentheses, for example "Casein (milk)," or as an alternative, there must be a statement separate but adjacent to the ingredients list: "Contains milk" (and any other of the allergens with mandatory labeling). Dairy-sourced protein ingredients include casein, caseinates, whey and lactalbumin, among others. The U.S. FDA has a recall process for foods that contain undeclared allergenic ingredients. The University of Wisconsin has a list of foods that may contain dairy proteins, yet are not always obvious from the name or type of food. This list contains the following examples:

- Bread, baked goods and desserts,
- Caramel and nougat candies,
- Cereals, crackers, food bars,
- Chewing gum,
- Chocolate ('milk' and 'dark'),
- "Cream of" soups,
- Creamy pasta sauces,
- Creamy salad dressings,
- Eggnog,
- Flavored potato chips,
- Hot dogs and lunch meat,
- Instant mashed potatoes,
- Margarine,
- Medical food beverages,
- Non-dairy creamer,
- Sherbet,
- Pudding and custard.

There is a distinction between "Contains" and "May contain". The first is a deliberate addition to the ingredients of a food, and is required. The second addresses unintentional possible inclusion of ingredients, in this instance dairy-sourced, during transportation, storage or at the manufacturing site, and is voluntary, and is referred to as precautionary allergen labeling (PAL).

Milk from other mammalian species (goat, sheep, etc.) should not be used as a substitute for cow's milk, as milk proteins from other mammals are often cross-reactive. Nevertheless, some people with cow's milk allergy can tolerate goat's or sheep's milk, and vice versa. Milk from camels, pigs, reindeer, horses, and donkeys may also be tolerated in some cases. Probiotic products have been tested, and some found to contain milk proteins which were not always indicated on the labels.

Cross-reactivity with Soy

Infants – either still 100% breastfeeding or on infant formula – and also young children may be prone to a combined cow's milk and soy protein allergy, referred to as "milk soy protein intolerance" (MSPI). A U.S. state government website presents the concept, including a recommendation that nursing mothers discontinue eating any foods that contain dairy or soy ingredients. In opposition to this recommendation, a published scientific review stated that there was not yet sufficient evidence in the human trial literature to conclude that maternal dietary food avoidance during lactation would prevent or treat allergic symptoms in breastfed infants.

A review presented information on milk allergy, soy allergy and cross-reactivity between the two. Milk allergy was described as occurring in 2.2% to 2.8% of infants and declining with age. Soy allergy was described as occurring in zero to 0.7% of young children. According to several studies cited in the review, between 10% and 14% of infants and young children with confirmed cow's milk allergy were determined to also be sensitized to soy and in some instances have a clinical reaction after consuming a soy-containing food. The research did not address whether the cause was two separate allergies or a cross-reaction due to a similarity in protein structure, as which occurs for cow's milk and goat's milk. Recommendations are that infants diagnosed as allergic to cow's milk infant formula be switched to an extensively hydrolyzed protein formula rather than a soy whole protein formula.

Prognosis

Milk allergy typically presents in the first year of life. The majority of children outgrow milk allergy by the age of ten years. One large clinical trial reported resolutions of 19% by age 4 years, 42% by age 8 years, 64% by age 12 years, and 79% by 16 years. Children are often better able to tolerate milk as an ingredient in baked goods relative to liquid milk. Childhood predictors for adult-persistence are anaphylaxis, high milk-specific serum IgE, robust response to the skin prick test and absence of tolerance to milk-containing baked foods. Resolution was more likely if baseline serum IgE was lower, or if IgE-mediated allergy was absent so that all that was present was cell-mediated, non-IgE allergy. People with confirmed cow's milk allergy may also demonstrate an allergic response to beef, more so to rare beef versus well-cooked beef. The offending protein appears to be bovine serum albumin.

Milk allergy has consequences. In a U.S. government diet and health surveys conducted in 2007–2010, 6,189 children ages 2–17 years were assessed. For those classified as cow's milk allergic at the time of the survey, mean weight, height and body-mass index were significantly lower than their non-allergic peers. This was not true for children with other food allergies. Diet assessment showed a significant 23% reduction of calcium intake and near-significant trends for lower vitamin D and total calorie intake.

Soy Allergy

Blocks of tofu for sale.

Soy allergy is a type of food allergy. It is a hypersensitivity to dietary substances from soy, causing an overreaction of the immune system which may lead to severe physical symptoms for millions of people. The Asthma and Allergy Foundation of America estimates soy is among the eight most common food allergens for pediatric and adult food allergy patients. It is usually treated with an exclusion diet and vigilant avoidance of foods that may be contaminated with soy ingredients. The most severe food allergy reaction is called anaphylaxis and is a medical emergency requiring immediate attention and treatment with epinephrine.

Signs and Symptoms

Food allergies can have fast onset (from seconds to one hour) or slow onset (from hours to several days) depending on mechanism. Symptoms may include: rash, hives, itching of mouth, lips, tongue, throat, eyes, skin, or other areas, swelling of lips, tongue, eyelids, or the whole face, difficulty swallowing, runny or congested nose, hoarse voice, wheezing, shortness of breath, diarrhea, abdominal pain, lightheadedness, fainting, nausea and vomiting. Symptoms of allergies vary from person to person and may vary from incident to incident. Serious danger regarding allergies can begin when the respiratory tract or blood circulation is affected. The former can be indicated by wheezing, a blocked airway and cyanosis, the latter by weak pulse, pale skin, and fainting. When these symptoms occur the allergic reaction is called anaphylaxis. Anaphylaxis occurs when IgE antibodies are involved, and areas of the body that are not in direct contact with the food become affected and show severe symptoms. Untreated, this can proceed to vasodilation, a low blood pressure situation called anaphylactic shock, and death (very rare).

Non-IgE mediated reactions are slower to appear, and tend to manifest as gastrointestinal symptoms, without cutaneous or respiratory symptoms. Within non-IgE reactions, clinicians distinguish among food protein-induced enterocolitis syndrome (FPIES), food protein-induced allergic proctocolitis (FPIAP) and food protein-induced enteropathy (FPE). Common trigger foods for all are soy infant formula, and also cow's milk formula. FPIAP is considered to be at the milder end of the spectrum, and is characterized by intermittent bloody stools. FPE is identified by chronic diarrhea which will resolve when the offending food is removed from the infant's diet. FPIES can be severe, characterized by persistent vomiting 1–4 hours after an allergen-containing food, to the point of lethargy. Watery and sometimes bloody diarrhea can develop 5–10 hours after the triggering meal, to the point of dehydration and low blood pressure. Infants reacting to soy formula may also react to cow's milk formula. International consensus guidelines have been established for the diagnosis and treatment of FPIES.

Causes

Sources of Soy Protein

Many fast-food restaurants commonly use soy protein in hamburger buns (soy flour), hamburger meat (soy protein) and hydrolyzed vegetable protein (HVP) in sauces. On their respective websites, McDonald's and Burger King list soy flour as an ingredient in their hamburger buns. Multigrain breads, doughnuts, doughnut mix and pancake mix commonly contain soy flour. Ice cream can also contain soy ingredients. Canned tuna may contain vegetable broth which contains soy protein. Some food contains soy-based ingredients that are not considered allergens under national regulations, and thus not labeled, for example, foods cooked in highly refined soy oil, which is considered safe due to absence of soy protein.

Products containing soy protein include:

- Edamame,
- Miso,
- Nattō,
- Shoyu sauce,
- Soy (soy albumin, soy fiber, soy flour, soy grits, soy milk, soy nuts, soy sprouts),
- Soybean (curd, granules),
- Soybean butter,
- Soy protein (concentrate, isolate),
- Soy milk,
- Soy sauce, tamari,
- Tempeh,
- Textured vegetable protein (TVP),
- Hydrolyzed vegetable protein (HVP),
- Tofu.

The following food additives may contain soy protein:

- Flavoring (including natural and artificial).
- Prepared broths, including chicken broth, vegetable broth, and bouillon cubes.

Non-allergenic Soy Derivatives

Highly refined soybean oil contains very little amounts of soy proteins, and is usually safe for people with soy allergy. They are exempt from being labelled as a major allergen under US FDA regulations. Note that soybean oils extracted using mechanical means (pressed or extruded) should be avoided.

Soy lecithin, usually derived from soybean oil, likewise presents minimal amounts of soy protein. However, the US FDA only exempts a few soy lecithin products from its mandatory allergenic source labeling requirements.

Cross-reactivity with Dairy

Infants - either still 100% breastfeeding or partially/entirely on infant formula - and also young children - may be prone to a combined cow's milk and soy protein allergy referred to as "milk soy protein intolerance" (MSPI). A US state government website presents the concept, including a recommendation that nursing mothers discontinue eating any foods that contain dairy or soy ingredients. In opposition to this recommendation, a published scientific review stated that there was not yet sufficient evidence in the human trial literature to conclude that maternal dietary food avoidance during lactation would prevent or treat allergic symptoms in breastfed infants.

A review presented information on soy allergy, milk allergy and cross reactivity between the two. Milk allergy was described as occurring in 2.2% to 2.8% of infants and declining with age. Soy allergy was described as occurring in zero to 0.7% of young children. According to several studies cited in the review, between 10% and 14% of infants and young children with confirmed cow's milk allergy were determined to also be sensitized to soy and in some instances have a clinical reaction after consuming a soy-containing food. The research did not address whether the cause was two separate allergies or a cross-reaction due to a similarity in protein structure, as which occurs for cow's milk and goat's milk. Recommendations are that infants diagnosed as allergic to cow's milk infant formula be switched to an extensively hydrolyzed protein formula rather than a soy whole protein formula.

Dosage Tolerance

Many people with soy allergy can tolerate small or moderate amounts of soy protein: the typical dose needed to induce an allergic response is about 100 times higher than for many other food allergens.

Mechanisms

Conditions caused by food allergies are classified into three groups according to the mechanism of the allergic response:

- IgE-mediated (classic): The most common type, manifesting acute changes that occur shortly after eating, and may progress to anaphylaxis.

- Non-IgE mediated: Characterized by an immune response not involving immunoglobulin E; may occur hours to days after eating, complicating diagnosis.
- IgE and non-IgE-mediated: A hybrid of the above two types.

Allergic reactions are hyperactive responses of the immune system to generally innocuous substances, such as proteins in the foods we eat. Why some proteins trigger allergic reactions while others do is not entirely clear, although in part thought to be due to resistance to digestion. Because of this, intact or largely intact proteins reach the small intestine, which has a large presence of white blood cells involved in immune reactions. The heat of cooking can help make protein molecules less allergenic.

In the early stages of acute allergic reaction, lymphocytes previously sensitized to a specific protein or protein fraction react by quickly producing a particular type of antibody known as secreted IgE (sIgE), which circulates in the blood and binds to IgE-specific receptors on the surface of other kinds of immune cells called mast cells and basophils. Activated mast cells and basophils undergo a process called degranulation, during which they release histamine and other inflammatory chemical mediators into the surrounding tissue causing effects, such as vasodilation, mucous secretion, nerve stimulation, and smooth-muscle contraction. This results in runny nose, itchiness, shortness of breath, and potentially anaphylaxis. Depending on the individual, the allergen, and the mode of introduction, the symptoms can be system-wide (classical anaphylaxis), or localized to the respiratory system (asthma) or skin (eczema). After the chemical mediators of the acute response subside, typically 2–24 hours after the original reaction, late-phase responses referred to as non-IgE mediated can occur due to the migration of other types of white blood cells to the initial reaction sites.

Allergenic proteins from soy (*Glycine max*) are named under a nomenclature decided by IUIC, which is also responsible for numbering many of the proteins. Proteins numbered by IUIC include:

- Gly m 1, a hydrophobic protein;
- Gly m 2, defensin;
- Gly m 3, profilin;
- Gly m 4, PR-10;
- Gly m 5, vicilin, a cupin;
- Gly m 6, legumin, a cupin;
- Gly m 7, seed biotinylated protein;
- Gly m 8, 2S albumin.

These proteins are recognized by the immune system as antigens in susceptible individuals. As many as 8 other soy allergenic proteins has been reported in literature. Peanut, a legume, shares many similar proteins with soy, so individuals with soy allergy may have peanut allergy too.

Diagnosis

Diagnosis of soy allergy is based on the person's history of allergic reactions, skin prick test (SPT), patch test and measurement of soy protein specific serum immunoglobulin E (IgE or sIgE). A negative IgE test does not rule out non-IgE mediated allergy, also described as cell-mediated allergy. SPT and sIgE have sensitivities of 55% and 83% respectively, and specificities of 68% and 38%. These numbers mean that either test may miss diagnosing an existing soy allergy, and that both can also be positive for other food allergens. Confirmation is by double-blind, placebo-controlled food challenges, conducted by an allergy specialist.

Treatment

Treatment for accidental ingestion of soy products by allergic individuals varies depending on the sensitivity of the person. An antihistamine such as diphenhydramine (Benadryl) may be prescribed. Sometimes prednisone will be prescribed to prevent a possible late phase Type I hypersensitivity reaction. Severe allergic reactions (anaphalaxis) may require treatment with an epinephrine pen, i.e., an injection device designed to be used by a non-healthcare professional when emergency treatment is warranted. A second dose is needed in 16-35% of episodes.

Wheat Allergy

Wheat allergy is an allergy to wheat which typically presents itself as a food allergy, but can also be a contact allergy resulting from occupational exposure. Like all allergies, wheat allergy involves immunoglobulin E and mast cell response. Typically the allergy is limited to the seed storage proteins of wheat. Some reactions are restricted to wheat proteins, while others can react across many varieties of seeds and other plant tissues. Wheat allergy is rare. Prevalence in adults was found to be 0.21% in a 2012 study in Japan.

Wheat allergy may be a misnomer since there are many allergenic components in wheat, for example serine protease inhibitors, glutelins and prolamins and different responses are often attributed to different proteins. Twenty-seven potential wheat allergens have been successfully identified. The most severe response is exercise/aspirin induced anaphylaxis attributed to one omega gliadin that is a relative of the protein that causes celiac disease. Other more common symptoms include nausea, urticaria, and atopy.

Gluten sensitivity is not usually classified as a wheat allergy. Management of wheat allergy consists of complete withdrawal of any food containing wheat and other gluten-containing cereals (gluten-free diet).

Types of Allergens

There are four major classes of seed storage proteins: albumins, globulins, prolamins and glutelins. Within wheat, prolamins are called gliadins and glutelins are called glutenins. These two protein groups form the classic glutens. While gluten is also the causative agent of celiac disease (CD), celiac disease can be contrasted to gluten allergy by the involvement of different immune cells and antibody types, and because the list of allergens extend beyond the classic gluten category of proteins.

Gluten Allergy

Prolamin Allergies

Prolamins and the closely related glutelins, a recent study in Japan found that glutenins are a more frequent allergen, however gliadins are associated with the most severe disease. A proteomics based study found a γ-gliadin isoform gene.

Glutelin Allergies

Glutenin (wheat glutelin) is a predominant allergen in wheat. Nine subunits of LMW-glutenin have been linked in connection with wheat allergies.

Albumin and Globulin Allergy

At present many of the allergens of wheat have not been characterized; however, the early studies found many to be in the albumin class. A recent study in Europe confirmed the increased presence of allergies to amylase/trypsin inhibitors (serpins) and lipid transfer protein (LPT), but less reactivity to the globulin fraction. The allergies tend to differ between populations (Italian, Japanese, Danish or Swiss), indicating a potential genetic component to these reactivities.

Other Allergies

Wheat Pollen and Grass Allergies

Respiratory allergies are an occupational disease that develop in food service workers. Previous studies detected 40 allergens from wheat; some cross-reacted with rye proteins and a few cross-reacted with grass pollens. A later study showed that baker's allergy extend over a broad range of cereal grasses (wheat, durum wheat, triticale, cereal rye, barley, rye grass, oats, canary grass, rice, maize, sorghum and Johnson grass) though the greatest similarities were seen between wheat and rye, and that these allergies show cross reactivity between seed proteins and pollen proteins, including a prominent crossreactivity between the common environment rye pollen and wheat gluten.

Derivative Allergies

Proteins are made of a chain of dehydrated amino acids. When enzymes cut proteins into pieces they add water back to the site at which they cut, called enzymatic hydrolysis, for proteins it is called proteolysis. The initial products of this hydrolysis are polypeptides, and smaller products are called simply peptides; these are called *wheat protein hydrolysates*. These hydrolysates can create allergens out of wheat proteins that previously did not exist by the exposure of buried antigenic sites in the proteins.

When proteins are cut into polypeptides, buried regions are exposed to the surface, and these buried regions may possibly be antigenic. Such hydrolyzed wheat protein is used as an additive in foods and cosmetics. The peptides are often 1 kD in size (9 amino acid residues in length) and may increase the allergic response. These wheat polypeptides can cause immediate contact urticaria in susceptible people.

Signs and Symptoms

Wheat allergies are not altogether different from other food allergies or respiratory allergies. However two conditions, exercise/aspirin induced anaphylaxis and urticaria, occur more frequently with wheat allergies.

Common symptoms of a wheat allergy include sacroiliitis, eczema (atopic dermatitis), hives (urticaria), asthma, "hay fever" (allergic rhinitis), angioedema (tissue swelling due to fluid leakage from blood vessels), abdominal cramps, nausea, and vomiting. Rarer symptoms include anaphylactic shock, anxiety, arthritis, bloated stomach, chest pains, depression or mood swings, diarrhea, dizziness, headache, joint and muscle aches and pains (may be associated with progressive arthritis), palpitations, psoriasis, irritable bowel syndrome (IBS), swollen throat or tongue, tiredness and lethargy, and unexplained cough.

Asthma, Anaphylaxis and Nasal Allergies

Exercise-induced Anaphylaxis

Wheat gliadins and potentially oat avenins are associated with another disease, known as wheat-dependent exercise induced anaphylaxis (WDEIA) which is similar to baker's allergy as both are mediated by IgE responses. In WDEIA, however, the ω-gliadins or a high molecular weight glutenin subunit, and similar proteins in other *Triticeae* genera, enter the blood stream during exercise where they cause acute asthmatic or allergic reaction. Wheat may specifically induce WDEIA and certain chronic urticaria because the anti-gliadin IgE detects ω_5-gliadins expressed by most of the Gli-B1 alleles, but prolamins extracted from rye or wheat/rye translocates invoke almost no responses. The Gli-B1 gene in wheat, *Triticum aestivum*, comes from the progenitor species *Aegilops speltoides*. This indicates that nascent mutations on the B genome of wheat are from a small number of cultivated Triticeae species.

Baker's Allergy

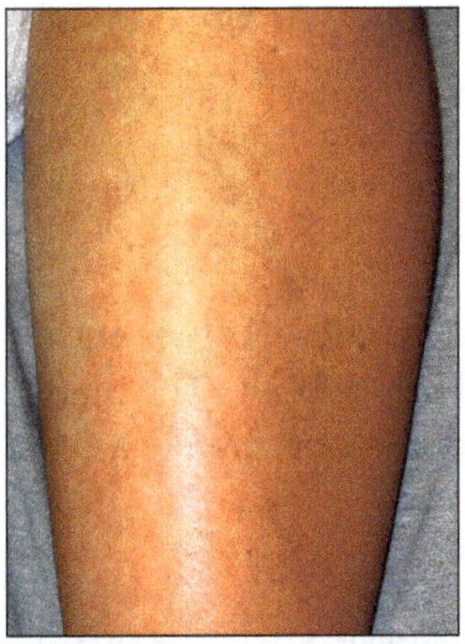

Allergic urticaria on the shin.

Baker's allergy has a ω-gliadin component and thioredoxin hB component. In addition, a gluten-extrinsic allergen has been identified as aspergillus amylase, added to flour to increase its baking properties.

Urticaria, Atopy and Eczema

Contact sensitivity, atopic dermatitis, eczema, and urticaria appear to be related phenomena, the cause of which is generally believed to be the hydrophobic prolamin components of certain Triticeae, Aveneae cultivars. In wheat one of these proteins is ω-gliadin (Gli-B1 gene product). A study of mothers and infants on an allergen-free diet demonstrated that these conditions can be avoided if wheat sensitive cohort in the population avoid wheat in the first year of life. As with exercise induced anaphylaxis, aspirin (also: tartrazine, sodium benzoate, sodium glutamate (MSG), sodium metabisulfite, tyramine) may be sensitizing factors for reactivity. Studies of the wheat-dependent exercise induced anaphylaxis demonstrate that atopy and EIA can be triggered from the ingestion of that aspirin and probably NSAIDs allow the entry of wheat proteins into the blood, where IgE reacts within allergens in the dermal tissues. Some individuals may be so sensitive that low dose aspirin therapy can increase risk for both atopy and WDEIA.

Wheat allergies were also common with contact dermatitis. A primary cause was the donning agent used for latex gloves prior to the 1990s, however most gloves now use protein free starch as a donning agent.

Rheumatoid Arthritis

There appears to be an association of rheumatoid arthritis (RA) both with gluten sensitive enteropathy (GSE) and gluten allergies. RA in GSE/CD may be secondary to tissue transglutaminase (tTG) autoimmunity. In a recent study in Turkey, 8 of 20 RA patients had wheat reactivities on the radioallergosorbent test (RAST). When this allergic food and all other patient specific RAST+ foods were removed half of the patients had improved RA by serological markers. In patients with wheat allergies, rye was effectively substituted. This may indicate that some proportion of RA in GSE/CD is due to downstream effects of allergic responses. In addition, cross-reactive anti-beef-collagen antibodies (IgG) may explain some rheumatoid arthritis (RA) incidences.

Neuropathies

- Migraines: In the late 70s it was reported that people with migraines had reactions to food allergens, like RA, the most common reaction was to wheat (78%), orange, eggs, tea, coffee, chocolate, milk, beef, corn, cane sugar, and yeast. When 10 foods causing the most reactions were removed migraines fell precipitously, hypertension declined. Some specific instances are attributed to wheat.
- Autism: Parents of children with autism often ascribe the children's gastrointestinal symptoms to allergies to wheat and other foods. The published data on this approach are sparse, with the only double-blind study reporting negative results.

Diagnosis

Diagnoses of wheat allergy may deserve special consideration. Omega-5 gliadin, the most potent wheat allergen, cannot be detected in whole wheat preparations; it must be extracted and partially digested (similar to how it degrades in the intestine) to reach full activity. Other studies show that digestion of wheat proteins to about 10 amino acids can increase the allergic response 10 fold. Certain allergy tests may not be suitable to detect all wheat allergies, resulting in cryptic allergies. Because many of the symptoms associated with wheat allergies, such as sacroiliitis, eczema and asthma, may be related or unrelated to a wheat allergy, medical deduction can be an effective way of determining the cause. If symptoms are alleviated by immunosuppressant drugs, such as Prednisone, an allergy-related cause is likely. If multiple symptoms associated with wheat allergies are present in the absence of immunosuppressants then a wheat allergy is probable.

Prevention and Treatment

Management of wheat allergy consists of complete withdrawal of any food containing wheat and other gluten-containing cereals (gluten-free diet). Nevertheless, some patients can tolerate barley, rye or oats.

In people suffering less severe forms of wheat-dependent exercise induced anaphylaxis (WDEIA), may be enough completely avoiding wheat consumption before exercise and other cofactors that trigger disease symptoms, such as nonsteroidal anti-inflammatory drugs and alcohol.

Wheat is often a cryptic contaminant of many foods; more obvious items are bread crumbs, maltodextrin, bran, cereal extract, couscous, cracker meal, enriched flour, gluten, high-gluten flour, high-protein flour, seitan, semolina wheat, vital gluten, wheat bran, wheat germ, wheat gluten, wheat malt, wheat starch or whole wheat flour. Less obvious sources of wheat could be gelatinized starch, hydrolyzed vegetable protein, modified food starch, modified starch, natural flavoring, soy sauce, soy bean paste, hoisin sauce, starch, vegetable gum, specifically beta-glucan, vegetable starch.

Alternative Cereals

Triticeae gluten-free oats (free of wheat, rye or barley) may be a useful source of cereal fiber. Some wheat allergies allow the use of rye bread as a substitute. Rice flour is a commonly used alternative for those allergic to wheat. Wheat-free millet flour, buckwheat, flax seed meal, corn meal, quinoa flour, chia seed flour, tapioca starch or flour, and others can be used as substitutes.

Peanut Allergy

Peanut allergy is one of the most common causes of severe allergy attacks. Peanut allergy symptoms can be life-threatening (anaphylaxis). For some people with peanut allergy, even tiny amounts of peanuts can cause a serious reaction.

Peanut allergy has been increasing in children. Even if you or your child has had only a mild allergic reaction to peanuts, it's important to talk to your doctor. There is still a risk of a more serious future reaction.

Symptoms

An allergic response to peanuts usually occurs within minutes after exposure. Peanut allergy signs and symptoms can include:

- Runny nose.
- Skin reactions, such as hives, redness or swelling.
- Itching or tingling in or around the mouth and throat.
- Digestive problems, such as diarrhea, stomach cramps, nausea or vomiting.
- Tightening of the throat.
- Shortness of breath or wheezing.

Anaphylaxis: A Life-threatening Reaction

Peanut allergy is the most common cause of food-induced anaphylaxis, a medical emergency that requires treatment with an epinephrine (adrenaline) injector (EpiPen, Symjepi, others) and a trip to the emergency room.

Anaphylaxis signs and symptoms can include:

- Constriction of airways.
- Swelling of the throat that makes it difficult to breathe.
- A severe drop in blood pressure (shock).
- Rapid pulse.
- Dizziness, lightheadedness or loss of consciousness.

Causes

Peanut allergy occurs when your immune system mistakenly identifies peanut proteins as something harmful. Direct or indirect contact with peanuts causes your immune system to release symptom-causing chemicals into your bloodstream.

Exposure to peanuts can occur in various ways:

- Direct contact: The most common cause of peanut allergy is eating peanuts or peanut-containing foods. Sometimes direct skin contact with peanuts can trigger an allergic reaction.
- Cross-contact: This is the unintended introduction of peanuts into a product. It's generally the result of a food being exposed to peanuts during processing or handling.
- Inhalation: An allergic reaction may occur if you inhale dust or aerosols containing peanuts, from a source such as peanut flour or peanut oil cooking spray.

Risk Factors

It isn't clear why some people develop allergies while others don't. However, people with certain risk factors have a greater chance of developing peanut allergy.

Peanut allergy risk factors include:

- Age: Food allergies are most common in children, especially toddlers and infants. As you grow older, your digestive system matures, and your body is less likely to react to food that triggers allergies.
- Past allergy to peanuts: Some children with peanut allergy outgrow it. However, even if you seem to have outgrown peanut allergy, it may recur.
- Other allergies: If you're already allergic to one food, you may be at increased risk of becoming allergic to another. Likewise, having another type of allergy, such as hay fever, increases your risk of having a food allergy.
- Family members with allergies: You're at increased risk of peanut allergy if other allergies, especially other types of food allergies, are common in your family.
- Atopic dermatitis: Some people with the skin condition atopic dermatitis (eczema) also have a food allergy.

References

- Urisu A, Kondo Y, Tsuge I (2015). Hen›s Egg Allergy. Chem Immunol Allergy. Chemical Immunology and Allergy. 101. Pp. 124–30. Doi:10.1159/000375416. ISBN 978-3-318-02340-4. PMID 26022872
- Corn-allergy: verywellhealth.com, Retrieved 8 January, 2019
- Caubet JC, Wang J (2011). «Current understanding of egg allergy». Pediatr. Clin. North Am. 58 (2): 427–43, xi. Doi:10.1016/j.pcl.2011.02.014. PMC 3069662. PMID 21453811
- Allergies and Food Safety». United States Department of Agriculture. Food Safety and Inspection Service. 1 December 2016. Retrieved 16 February 2018
- Holt PG, Sly PD (October 2007). «Th2 cytokines in the asthma late-phase response». Lancet. 370 (9596): 1396–8. Doi:10.1016/S0140-6736(07)61587-6. PMID 17950849
- Janeway C, Paul Travers, Mark Walport, Mark Shlomchik (2001). Immunobiology; Fifth Edition. New York and London: Garland Science. Pp. E–book. ISBN 978-0-8153-4101-7
- Symptoms-causes, peanut-allergy, diseases-conditions: mayoclinic.org, Retrieved 13 May, 2019

Permissions

All chapters in this book are published with permission under the Creative Commons Attribution Share Alike License or equivalent. Every chapter published in this book has been scrutinized by our experts. Their significance has been extensively debated. The topics covered herein carry significant information for a comprehensive understanding. They may even be implemented as practical applications or may be referred to as a beginning point for further studies.

We would like to thank the editorial team for lending their expertise to make the book truly unique. They have played a crucial role in the development of this book. Without their invaluable contributions this book wouldn't have been possible. They have made vital efforts to compile up to date information on the varied aspects of this subject to make this book a valuable addition to the collection of many professionals and students.

This book was conceptualized with the vision of imparting up-to-date and integrated information in this field. To ensure the same, a matchless editorial board was set up. Every individual on the board went through rigorous rounds of assessment to prove their worth. After which they invested a large part of their time researching and compiling the most relevant data for our readers.

The editorial board has been involved in producing this book since its inception. They have spent rigorous hours researching and exploring the diverse topics which have resulted in the successful publishing of this book. They have passed on their knowledge of decades through this book. To expedite this challenging task, the publisher supported the team at every step. A small team of assistant editors was also appointed to further simplify the editing procedure and attain best results for the readers.

Apart from the editorial board, the designing team has also invested a significant amount of their time in understanding the subject and creating the most relevant covers. They scrutinized every image to scout for the most suitable representation of the subject and create an appropriate cover for the book.

The publishing team has been an ardent support to the editorial, designing and production team. Their endless efforts to recruit the best for this project, has resulted in the accomplishment of this book. They are a veteran in the field of academics and their pool of knowledge is as vast as their experience in printing. Their expertise and guidance has proved useful at every step. Their uncompromising quality standards have made this book an exceptional effort. Their encouragement from time to time has been an inspiration for everyone.

The publisher and the editorial board hope that this book will prove to be a valuable piece of knowledge for students, practitioners and scholars across the globe.

Index

www.ingramcontent.com/pod-product-compliance
Lightning Source LLC
LaVergne TN
LVHW080648210926
848485LV00003B/348
9781641726740